WITHDRAWN
UTSA Libraries

United States and Britain in
Diego Garcia

United States and Britain in Diego Garcia

The Future of a Controversial Base

Peter H. Sand

UNITED STATES AND BRITAIN IN DIEGO GARCIA
Copyright © Peter H. Sand, 2009.

All rights reserved.

First published in 2009 by
PALGRAVE MACMILLAN®
in the United States—a division of St. Martin's Press LLC,
175 Fifth Avenue, New York, NY 10010.

Where this book is distributed in the UK, Europe and the rest of the world, this is by Palgrave Macmillan, a division of Macmillan Publishers Limited, registered in England, company number 785998, of Houndmills, Basingstoke, Hampshire RG21 6XS.

Palgrave Macmillan is the global academic imprint of the above companies and has companies and representatives throughout the world.

Palgrave® and Macmillan® are registered trademarks in the United States, the United Kingdom, Europe and other countries.

ISBN: 978-0-230-61709-4

Library of Congress Cataloging-in-Publication Data is available from the Library of Congress.

A catalogue record of the book is available from the British Library.

Design by Newgen Imaging Systems (P) Ltd., Chennai, India.

First edition: July 2009

10 9 8 7 6 5 4 3 2 1

Printed in the United States of America.

CONTENTS

List of Illustrations		vii
Preface		ix
1	History: Empire's Last-Born Colony	1
2	Human Rights: How to Depopulate an Island	15
3	Power Politics: "Our" Ocean	35
4	Military Secrecy: Public Access Denied	43
5	Nemesis: Natural Heritage Dredged—and Drowned	51
6	Epilogue: The Lords' Day?	63
Appendices (U.S.-UK Official Documents on Diego Garcia)		69
I	Agreement London, December 1966	69
II	Agreed Confidential Minutes, London, December 1966	81
III	Supplement London, October 1972	84
IV	Supplement London, February 1976	91
V	Amendment London, June 1976	105
VI	Supplement Washington, D.C., December 1982	107
VII	Supplement Washington, D.C., November 1987	109
VIII	Agreement London, June/July 1999	111

IX	Exchanges of Letters, London–Washington, D.C., 2001–2004	116
X	House of Lords Judgment, London, October 2008	122

Notes	145
Index of Names	193
Subject Index	203

ILLUSTRATIONS

Maps

1 Indian Ocean Map—Diego Garcia, 2007   x
2 U.S. Navy, Existing Land-Use Map Diego Garcia, 1997   12
3 Ramsar Convention Map—Diego Garcia Protected Site, 2001   61

Exhibits

1 UK-U.S. Secret Side-Note to Agreement on Diego Garcia, 1966   6
2 UK Foreign and Commonwealth Office, Resettlement Memorandum, 1971   18
3 UK-Mauritius Îlois Claims Agreement, 1982   26

Figure

1 IRIS/IDA Seismic Data Diego Garcia, *Tsunami,* 2004   46

Table

1 U.S. Congress, Diego Garcia Construction Budget 1970–1987   36

PREFACE

> *Demain Diego:*
> *pa rakont zistwar*
> *abriti nou ar lespwar.*
> *Demain Diego:*
> *anglé mérikin mantèr*
> *kouraz anvolé ar volèr.*
> MICHEL DUCASSE[1]

This book is intended as both primer and template—to tell the exemplary story of a "strategic island" base from a variety of angles: colonial history, and the plight of the islanders; power politics, and the role of the military; and finally, perhaps least understood, the lasting human impact of the base on a once-pristine environment.

Diego Garcia/British Indian Ocean Territory, arguably one of the most important (and certainly one of the most expensive) American bases overseas, has remained virtually "unknown in the US"[2]—as in the rest of the world, for that matter. Yet, in the words of a military historian, "ignorance is a dangerous thing. When accompanied by arrogance, a religious conviction in the correctness of one's vision, and elitism, it can be fatal."[3]

The author gratefully acknowledges valuable assistance and information received from Michiko Baba, Carl Bruch, April A. Christensen, Darryl K. Creasy, Chloe Davies, Steven J. Forsberg, Richard Gifford, Sandra Hails, Simon Hughes, Heiner Igel, Stjepan Keckes, Heinz Kluge, Walter C. Ladwig III, Irve C. Le Moyne Jr., Dominique Loye, Lucy McAllister, Ted A. Morris, Jay Nelson, Andrew R.G. Price, Richard Rahm, Charles R.C. Sheppard, David R. Snoxell, Clive Stafford Smith, Karen Sumida, David Vine, and Joanne Yeadon. Comments by two anonymous reviewers of the manuscript and editorial guidance by Farideh Koohi-Kamali and Asa Johnson at Palgrave Macmillan were highly appreciated. Any views and opinions expressed here are, however, the sole responsibility of the author.

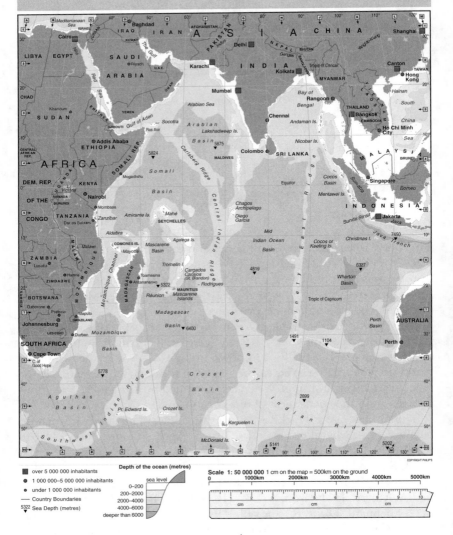

Map 1 Indian Ocean Map—Diego Garcia, 2007[4]

Source: Courtesy of Éditions de l'Océan Indien, Mauritius.

CHAPTER 1

History: Empire's Last-Born Colony

Once upon a time there was a proverbial tropical island—palm trees on a coral atoll, in the middle of the Indian Ocean; white beaches; and a coconut plantation—inhabited by a few hundred Creole-speaking people who called themselves *Îlois* [ilwa] (the Islanders). The island was named *Diego García* after a Portuguese explorer in Spanish services, who claimed to have discovered it about 1532.[1]

Various foreigners arrived in ships from time to time and planted flags on the island: the French in 1769, the British in 1786, the French again in 1786, and the British again from 1810 onward.[2] In October 1914, the German gunboat *Emden* paid a surprise visit[3] and in 1942, the RAF set up a flying-boat base to guard against Japanese submarines; otherwise, the islanders were generally left in peace. The copra plantation, while not exactly flourishing, survived;[4] Jesuit missionaries built a chapel;[5] and the local fauna now increased to include some mules, dogs, chicken, and rats.

Then came the cold war, with Soviet naval operations expanding to the Indian Ocean, and the emergence of a Chinese threat. The U.S. Navy, too, began to look at strategic real estate in the region. It was Stuart B. Barber, the (civilian) assistant director of the navy's Long-Range Objectives Group (CNO/OP-93), established in 1955, who first targeted Diego Garcia as most suitable for his "strategic islands" concept. His argument was that the United States—in anticipation of the independence of former colonial territories in the Southern Hemisphere—should acquire base rights in certain strategically located islands so as to stockpile them for future use as communication, refuelling and "prepositioning" stations.[6]

A succession of American admirals and survey teams now arrived to inspect Diego Garcia as a prospective site: in 1957 (year of the Sputnik), Admiral Jerauld Wright, commander of the U.S. Atlantic Fleet; in 1961 (year of the Berlin Wall), Rear Admiral Jack Grantham, whose visit was accidentally filmed by missionaries unaware of the significance of their visitors;[7] and in 1964 (year of the Gulf of Tonkin incident, which led to full-scale U.S. involvement in Vietnam), a joint U.S.-UK military survey led by Commander Harry S. Hart from the office of the chief of U.S. naval operations.[8] After alternative locations (in particular, the tortoise island of Aldabra in the Seychelles) had to be turned down in the face of protests by nature conservationists and scientists (from the Royal Society of London and the Smithsonian Institution of Washington, D.C., in particular),[9] the Pentagon made up its mind that Diego Garcia was what it needed—with Admiral Horacio Rivero, Jr., vice chief of U.S. naval operations, blurting out at a briefing in the fall of 1964: "I want this island!"[10] So that is when the atoll made its entry on the stage of global politics and in modern world history.

From a strategic perspective, Diego Garcia's position was indeed unique. In the words of Admiral John S. McCain (grandfather of the unsuccessful 2008 presidential candidate), "as Malta is to the Mediterranean, Diego Garcia is to the Indian Ocean—equidistant from all points."[11] Moreover, the atoll's internal lagoon is a gigantic natural harbor—48 square miles wide (125 square kilometers), and 40–100 feet deep (12–30 meters)—protected on all sides by coral formations shaped like the sole of a foot when viewed from the air; hence its lofty American label, "footprint of freedom."[12]

In legal terms, however, the Chagos Archipelago to which the island geographically belongs was still part of the British Empire, administered by the Colonial Office through the royal governor of Mauritius.[13] And under Article 73 of the UN Charter, the United Kingdom was accountable for submitting regular reports on its administration of non-self-governing territories, and their progress toward self-government, to the United Nations' so-called Committee of Twenty-Four in New York.[14] In December 1960, the UN General Assembly had indeed unanimously passed (with the United States, the United Kingdom, and some other colonial powers abstaining) the Declaration on the Granting of Independence to Colonial Countries and People (Resolution 1514/XV), which provided for immediate steps to be taken in all "territories which have not yet attained independence, to transfer all powers to the peoples of those territories, without any conditions or reservations" and further cautioned that

> any attempt aimed at the partial or total disruption of the national unity and the territorial integrity of a country is incompatible with the purposes and principles of the Charter of the United Nations.[15]

Anglo-American negotiations on the future military uses of Diego Garcia thus faced a serious dilemma: how to separate the island from Mauritius, which had already been granted an intermediate level of self-government in 1964 and which was due to attain full independence in 1968? As a Colonial Office minister wrote on October 20, 1964,

> It would be unacceptable to both the British and the American defence authorities if facilities of the kind proposed were in any way to be subject to the political control of ministers of a newly emergent independent state.[16]

In April 1965, therefore, Colonial Secretary Anthony Greenwood (later Lord Greenwood of Rossendale) flew to Mauritius to propose a peculiar diplomatic arrangement:[17] the Chagos Archipelago was to be "excised" from the territory of Mauritius so as to be united with some of the neighboring outer islands of the British Seychelles colony (which in 1903 had been separated from Mauritius), henceforth to form a wholly *new* colonial territory distinct from either Mauritius or the Seychelles. In exchange, the two ex-colonies would upon independence receive a number of substantial counterpart benefits from the United Kingdom. In the case of Mauritius, those included £3 million ($8.4 million at the time) in cash, an Anglo-Mauritian defense funding agreement, and financial compensation for the expropriated plantation owners and for resettlement of the Chagossian *Îlois* in Mauritius.[18] In the case of the Seychelles, the United Kingdom agreed to build a new international airport at Mahé (for approximately £6 million, i.e., $16.8 million) and to buy out the private landowners on the two islands of Farquhar (who happened to be a Seychelles cabinet member and the principal shareholder of the Chagos plantation) and Desroches, for about half a million dollars each.[19]

During subsequent political negotiations held between top-level British and Mauritian representatives at London's Lancaster House in

September 1965, these terms were further discussed and orally agreed upon,[20] on the (unwritten) understanding that the Chagos islands would revert to Mauritius when they were no longer needed for defense purposes.[21] Accordingly, the UK government—without parliamentary debate—issued an Order-in-Council on November 8, 1965, proclaiming the formation of "a separate colony which shall be known as the British Indian Ocean Territory (BIOT)."[22] The new colony was to be administered from Victoria, the capital of the Seychelles, with the governor of the Seychelles acting as BIOT commissioner and chief magistrate. The order empowered the UK cabinet to govern the territory in classical colonial law fashion, that is, by simple executive "ordinances" issued by the commissioner, without parliamentary oversight and without participation by the resident population.[23]

International reactions to this fait accompli were less than enthusiastic. The BIOT "excision" was a blatant violation of UN Resolution 1514/XV—not only because it had been made a precondition for Mauritian independence (later described as "sheer blackmail" by a select committee of the Mauritian Legislative Assembly),[24] but also because it clearly disrupted the territorial integrity of two newly independent countries and hence infringed the general international legal rule of *uti possidetis*, which both the International Court of Justice and the Permanent Court of Arbitration at The Hague have invariably applied to ex-colonial boundary matters.[25] The UN General Assembly on December 16, 1965, thus passed Resolution 2066/XX—with 89 votes unopposed and 18 abstentions—calling on the United Kingdom "to take no action which would dismember the territory of Mauritius and violate its territorial integrity."[26]

Choosing to ignore both resolutions, however, the UK and U.S. governments went ahead to conclude a bilateral agreement, by exchange of notes, on the Availability for Defense Purposes of the British Indian Ocean Territory, in London on December 30,1966 (see appendices I–II),[27] followed by supplementary agreements and amendments also concluded by exchanges of notes in 1972, 1976, 1982, 1987, and 1999 (appendices III–VIII) and by subsequent exchanges of letters in 2001, 2002, and 2004 (appendix IX).[28] These agreements provide for the establishment and gradual upgrading of U.S. military facilities—under joint administration—on the island of Diego Garcia, successively expanded to include a naval communications facility, naval support facility, bomber forward operating location, and satellite tracking station; negotiations for a further upgrade supplement are currently under way.[29] The agreements expressly contemplate use of the base "for an indefinitely long period" and after an initial period of 50 years (i.e., in 2016) are to continue in force for a further period of 20 years unless terminated by two years' notice (appendix I, art. 11). The supplementary agreements of 1972 and 1976 specify that they "shall continue in force for as long as the BIOT Agreement continues in force or until such time as [the parties agree that] no part of Diego Garcia is any longer required for the purpose[s] of the facility, whichever occurs first" (appendix III, para. 20; appendix IV, para. 22).

After the Pentagon and Whitehall decided that the Diego Garcia facilities were adequate and alternative sites were no longer needed, the three western islands—Aldabra, Desroches, and Farquhar—were returned to the Seychelles in 1976 "as a gesture of goodwill" by the United Kingdom on the occasion of independence (and in the wake of repeated "condemnatory" requests to this effect by the UN Committee of Twenty-Four).[30] The 1965 Order-in-Council and the 1966 Diego Garcia Agreement were amended accordingly (appendix V), so that the BIOT currently consists of the Chagos Archipelago only (as shown on the colorful, redesigned BIOT postal stamps), governed by the director of the Overseas Territory Department at the UK Foreign and Commonwealth Office in London, as nonresident commissioner.[31]

While the UK-U.S. agreements contain no information on financial transactions between the two governments, the underlying deal is documented in a series of 1975 U.S. congressional hearings concerning the base.[32] In a secret *side-note* to the 1966 BIOT Agreement (Exhibit 1), the United States had agreed to provide up to half of the total "detachment costs" (i.e., including all bribes paid in Mauritius and the Seychelles), though not exceeding $14 million.[33] As the UK government was notoriously short of cash and did not wish to obtain parliamentary appropriations for this purpose, the U.S. Department of Defense established a secret trust fund under the 1963 POLARIS Sales Contract,[34] which had previously been negotiated pursuant to the Nassau Agreement between Prime Minister Harold Macmillan and President John F. Kennedy (December 18, 1962, for the supply of POLARIS nuclear ballistic missiles to Britain's submarines): The amount of $14 million was thus advanced as a "loan" to the United Kingdom, of which about $11.5 million were later waived as a 5 percent offset for "accrued research and development surcharges."[35] A subsequent investigation by the U.S. Government Accounting Office was unable to say whether these secret financial arrangements violated U.S. law, but concluded that the method of funding—a technique that masked real plans and costs—was clearly a circumvention of congressional oversight.[36]

Exhibit 1   UK-U.S. Secret Side-Note to Agreement on Diego Garcia, 1966[37]

I

Secret

Note No. 26   30 December 1966

From David K.E. Bruce,
U.S. Ambassador to London

To the Right Honorable George Brown, M.P.,
Secretary of State for Foreign Affairs,
Foreign Office, Whitehall, London, S.W.1

Sir,

I have the honor to refer to the Agreement concluded today between our two Governments concerning the availability of certain Indian Ocean islands for such defense needs of either of our two Governments as may arise. I wish to confirm the following financial arrangements which have been reached regarding the detachment of these islands from colonial administration and the acquisition of the lands thereon:

1. The United Kingdom will assume all costs pertaining to the administrative detachment of the Indian Ocean islands in question and to the acquisition of the lands thereon so that they may be available over the indefinite future to meet the defense needs of either Government as these needs arise.
2. Under the POLARIS Sales Agreement signed by our two Governments at Washington on April 6, 1963, the United Kingdom is obliged to make certain payments as a participation in expenditures incurred by the United States after January 1, 1963, for research and development of the POLARIS missile system (hereinafter "R&D surcharge"). Since the United Kingdom is assuming the costs of the administrative detachment of the Indian Ocean islands and of the acquisition of the lands thereon, the United States will forego the R&D surcharge to the extent of $14 million, or one half of the foregoing Indian Ocean islands costs incurred by the United Kingdom, whichever is the less. The amount of the R&D surcharge so foregone is referred to below as the contribution.

3. The procedure proposed for effecting the contribution is described in the following subparagraphs.
   (a) As of 30 September 1966 the United Kingdom has paid into the Trust Fund established pursuant to paragraph 2 of Article XI of the POLARIS Sales Agreement the aggregate sum of $14.3 million in respect of:
      (i) the R&D surcharge;
      (ii) the agreed overhead costs of the POLARIS program (hereinafter "Overhead"); and
      (iii) the agreed charge for use of all United States Government-furnished facilities (hereinafter "Facilities").
   Of this aggregate sum $14 million will be applied to meet current charges against the United Kingdom for the POLARIS procurement. In consequence, the next practicable quarterly payment by the United Kingdom into the Trust Fund for such current procurement charges will be reduced by the aforementioned amount of $14.0 million. If the next quarterly payment otherwise due for such current procurement charges is less than $14.0 million, the difference between the $14.0 million and the amount of that quarterly payment otherwise due will be deducted from the succeeding quarterly payment or payments made by the United Kingdom.
   (b) Beginning as of the last day of the quarter following the quarter in which this Exchange of Notes is signed and ending as of March 31, 1969, the United Kingdom will pay, in equal quarterly installments, the entire amounts for Overhead and Facilities called for by paragraph 2 of the Classified Minute relating to Article XI of the POLARIS Sales Agreement, less those amounts exceeding $14.0 million paid into the Trust Fund as of the date of signature of these arrangements in respect of the R&D surcharge and in respect of Overhead and Facilities, which balance will be applied against the first quarterly payment or payments for Overhead and Facilities.
   (c) When the cumulative amount of the R&D surcharge which would have been payable except for these arrangements equals the contribution, the United Kingdom will commence payments in respect of the R&D surcharge at the rate specified in subparagraph 1.b of Article XI of the POLARIS Sales Agreement. Should payments in respect of the R&D surcharge which would have been payable except for these arrangements prove insufficient to meet the contribution, the Governments of the United States and of the United Kingdom will consult in order to determine how the United States' obligation to provide the contribution can best be satisfied.

I have the honor to request you to confirm the foregoing financial arrangements on behalf of the Government of the United Kingdom.

Accept, Sir, the renewed assurances of my highest consideration.

[signed: *David Bruce*]
American Ambassador

II

Secret

No. AU 1199   30 December 1966

From the Minister of State, Foreign Office
London

To H.E. The Honourable David K.E. Bruce, CBE
Embassy of the United States of America, London

Your Excellency,

I have the honour to acknowledge receipt of your Note No. 26 of the 30th of December, 1966, concerning the financial arrangements which have been reached between the two Governments in respect of the detachment of the islands constituting the British Indian Ocean Territory from colonial administration and the acquisition of the lands thereon.

I have the honour to confirm on behalf of the Government of the United Kingdom the financial arrangements set out in Your Excellency's Note.

I have the honour to be, with the highest consideration, Your Excellency's obedient Servant,

(For the Secretary of State)
[signed: *Chalfont*]

Even though the Mauritian government had initially agreed to the Chagos "excision," it later alleged to have been misled and reclaimed sovereignty over the archipelago, by way of an amendment of the national Constitution (Article 111 of which now explicitly names "the Chagos Archipelago including Diego Garcia" as part of the national territory of Mauritius)[38] and in a series of statements to the United

Nations and other international assemblies.[39] The UK government invariably rejected the claim, while conceding that the islands would eventually be "ceded" to Mauritius at some unspecified future time "when they are no longer needed for defence purposes."[40] During top-level discussions in June 2008, however, prime ministers Gordon Brown and Dr. Navinchandra Ramgoolam agreed to place the Chagos sovereignty issue on the agenda of forthcoming intergovernmental consultations.[41]

The U.S. State Department, for its part, has consistently preferred to dodge the problem, by contending—in the words of Secretary of State Madeleine Albright—that "sovereignty issues are those between the United Kingdom and Mauritius."[42] That may no longer be good enough, though. Mauritius has begun to declare its membership in a growing number of multilateral treaties—such as the 1985 Eastern Africa Marine Environment Convention and its protocols—as extending to the Chagos Archipelago,[43] sometimes over protests by the United Kingdom (as in the case of the 1983 Strasbourg Convention on the Transfer of Sentenced Persons)[44] and sometimes in protest against similar extensions by the United Kingdom in turn (as in the case of the 1985/1987 UN treaties protecting the ozone layer).[45] This diplomatic can of worms will inevitably be reopened when the United States finally—for strategic reasons related to its territorial claims in the Arctic—ratifies the 1982 UN Convention on the Law of the Sea. The island of Diego Garcia now finds itself surrounded not only by a 200-mile zone claimed by the United Kingdom around the BIOT for "fisheries management and conservation" (since 1991) and for "environment protection and preservation" (since 2003),[46] but also by an overlapping 200-mile "exclusive economic zone" (EEZ) claimed by Mauritius around the same area since 1984 (formally recognized by the European Union since 1989[47] and reiterated upon signature of the 2006 Southern Indian Ocean Fisheries Agreement),[48] each legitimately declared under the Law of the Sea Convention.[49] To complicate matters further, the government of the Maldives has recently notified the UN Secretariat of its intention to challenge the legal boundaries of the BIOT's 200-mile zone, which overlap with its own EEZ claim.[50] (In contrast and contrary to an unfounded recent entry in the U.S. Central Intelligence Agency's *World Factbook*,[51] the Seychelles has never laid any territorial claims to the Chagos Archipelago.)[52]

Curiously enough, unlike the territorial waters of the metropolitan British Isles, which (like those of the United States) were extended from 3 to 12 nautical miles (22 kilometers) in the 1980s,[53] BIOT still has a *3-mile* territorial sea—except for the area around Diego Garcia, which the BIOT authorities have declared off-limits for all unauthorized vessels within *12 miles* from the shore.[54] Whereas the 200-mile zone now claimed by the United Kingdom around the Chagos Archipelago is thus contiguous to a *3-mile* zone of territorial waters,[55] the EEZ declared by Mauritius is contiguous to a *12-mile* territorial zone.[56] Moreover, the 2005 Mauritian Maritime Zones Regulations require prior authorization for the passage of "any ship carrying nuclear materials" through the internal, territorial, or archipelagic waters of Mauritius and require all foreign warships and submarines to obtain permission before transiting the country's EEZ (a requirement not recognized by the United States).[57]

These discrepancies and controversies hardly bode well for the international legal security of future American tenure in the archipelago, which is anything but "politically invulnerable."[58] Sooner or later, the U.S. Navy will have to bite the bullet and decide just whose maritime jurisdiction it recognizes over the waters encircling its base.[59] The question is not without ominous practical relevance. On the one hand, the United Kingdom formally extended protocols I and II (1977) of the Geneva Conventions relative to the Protection of Victims of Armed Conflicts—though *not* Geneva Conventions III and IV relative to the Treatment of Prisoners of War and to the Protection of Civilian Persons in Time of War (1949)[60]—to its BIOT in 2002.[61] On the other hand, Mauritius considers the 1966 International Covenants on Human Rights[62] as well as the 1984 UN Convention against Torture[63] applicable to Diego Garcia and its territorial waters, whereas the United Kingdom does *not*.[64] With regard to the International Criminal Court, Mauritius ratified the 1998 Rome Statute in 2002,[65] but then signed a bilateral immunity agreement with the United States—in exchange for commitments of technical/financial assistance—on June 25, 2003, exempting U.S. personnel on its territory from the jurisdiction of the court.[66]

Unlike Guantánamo Bay, however, Diego Garcia hardly lends itself to "black hole" hypotheses with regard to the legal status of the American base.[67] While the 1966 BIOT Agreement (appendix I below) grants U.S. military authorities the right to exercise criminal and disciplinary jurisdiction over persons subject to U.S. military law, the UK authorities retain "exclusive jurisdiction over members of the United States forces with respect to offences, including offences relating to security, punishable by law in force in the territory but *not* by the law of the United States."[68]

In practice, the jurisdiction of the United Kingdom in Diego Garcia is exercised by the resident representative of the BIOT commissioner (traditionally, a senior British naval officer), although a number of "gray areas" remain. For example, the United States is reported to keep substantial supplies of antipersonnel landmines in BIOT (e.g., some 10,000 mines in *Gator*, *Volcano*, and *MOPMS* dispenser packages for aircraft delivery),[69] the stockpiling and use of which are strictly prohibited by the 1997 Ottawa Landmine Convention (to which both the United Kingdom and Mauritius—though not the United States—are parties).[70] According to a statement by the UK Foreign Office,

> There are no US antipersonnel mines *on* Diego Garcia. We understand that the US stores munitions of various kinds on US warships anchored *off* Diego Garcia. Such vessels enjoy state immunity and are therefore outside the UK's jurisdiction and control. The US understands the importance we attach to their adherence to the Ottawa Convention as soon as possible.[71]

The UK representative at the Ottawa Convention's Standing Committee meeting in May 2003 even claimed that landmines on U.S. naval ships *inside British territorial waters* "are not on UK territory provided they remain on the ships."[72] This exoneration presumably would also exempt—without even a right of inspection—all stockpiled ordnance prohibited under the 1992 Chemical Weapons Convention (expressly extended to BIOT in 2005)[73] and the 2008 Cluster Munitions Convention.[74] It may be doubted, however, whether so generous an interpretation of international customary and treaty law will suffice to "black-hole" the entire fleet of American vessels in the Diego Garcia lagoon, that is, in British *internal* (rather than territorial) waters.

As is evident from the U.S. Navy's 1997 existing land-use map of the base (Map 2), all ammunition is stored either at depots and underground bunkers on land or on ships anchored in an "explosive safety arc" designated for this purpose *inside* the eastern part of the lagoon (Rambler Bay). After having spent millions of dollars to dredge that area for use by its "prepositioning" fleet, it would hardly make logistic sense for the United States to station ordnance cargoes in territorial

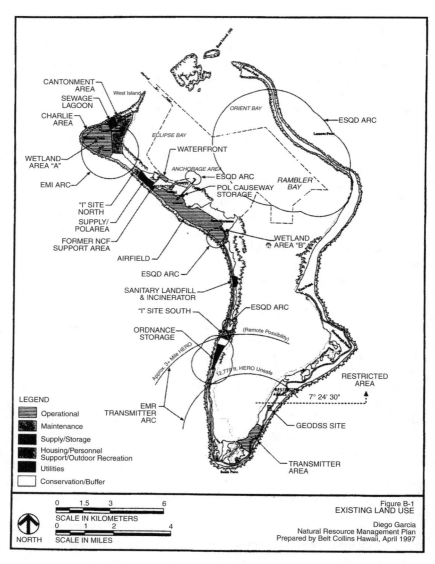

Map 2 U.S. Navy, Existing Land-Use Map Diego Garcia, 1997[75]

waters *outside* the protected lagoon—let alone at a distance of more than three nautical miles from the base.[76] At any rate, even "if anti-personnel mines were off-loaded onto land, e.g., to be transferred from ship to aircraft, this would not be consistent with our Ottawa Convention obligations," as noted by the chairman of the British Bar Human Rights Committee (BHRC) in a letter to the UK foreign secretary in November 2003.[77]

CHAPTER 2

Human Rights: How to Depopulate an Island

What made Diego Garcia so exceptionally attractive for military planners was a combination of its strategic location, isolation, and geophysical-meteorological features such as the relative scarcity of tropical storms.[1] There was a catch, though: unlike some other Indian Ocean atolls, the island had a human population—indeed settled for well over a century.[2] It is undisputed that at the time of the creation of the British Indian Ocean Territory (BIOT) in 1965, there were more than 1,500 people living on Diego Garcia and some of the other Chagos islands. While some of the inhabitants had migrated from Mauritius and the Seychelles as temporary workers with the copra plantations, a considerable number of indigenous families had their roots on Diego Garcia going back four or five generations, as evidenced by the tombstones preserved in the island cemetery.

Most of them had descended from slaves originally brought from Africa (mainly, Mozambique) by French colonists in the 18th century.[3] When the British Royal Navy conquered Diego Garcia in the Napoleonic Wars, they took many of the male African slaves off the island to the slave market in Goa where they were sold as "prizes"—that is, enemy property captured at sea—and eventually shipped to Sri Lanka to form the 4th Ceylon "Kaffir" Regiment, which fought in the Kandyan Expedition of 1815 and in the Uva Rebellion of 1818. Legal disputes over the proceeds from the booty, which netted £37 per head on average, continued in British courts well into the 1820s.[4] After slavery was abolished in Mauritius and its dependencies in 1834, the Îlois remaining in the

Chagos were "free, that is to say nominally, though perhaps very little change would be found in their condition."[5]

Yet, the prospective new tenants from the Pentagon had made it clear from the beginning that they were looking for an *uninhabited* site. Moreover, any inhabited territory would remain subject to Chapter XI of the UN Charter and hence to the reporting duties of Article 73(e) and ultimately to the universally declared goal of self-government.[6] The resident *Îlois* population thus posed a major risk of diplomatic embarrassment and precipitated some amazing creative thinking in London and Washington, D.C.

On February 25, 1966, the UK secretary of state for the colonies (Frank Pakenham, 7th Earl of Longford, KG PC) sent a confidential note to the newly appointed BIOT commissioner (Julian Asquith, 2nd Earl of Oxford and Asquith, KCMG), explaining that

> our primary objective in dealing with the people who are at present in the Territory must be to deal with them in the way which will best meet our future administrative and military needs and will at the same time ensure that they are given fair and just treatment... With these objectives in view we propose to avoid any reference to "permanent inhabitants," instead, to refer to the people in the islands as Mauritians and Seychellois... We are... taking steps to acquire ownership of the land on the islands and consider that it would be desirable... for the inhabitants to be given some form of temporary residence permit. We could then more effectively take the line in discussion that these people are Mauritian and Seychellois; that they are temporarily resident in BIOT for the purpose of making a living on the basis of contract or day to day employment with the companies engaged in exploiting the islands; and that when the new use of the islands makes it impossible for these operations to continue on the old scale the people concerned will be resettled in Mauritius or Seychelles.[7]

In a minute of June 1966, the BIOT commissioner noted accordingly that

> they [i.e., the Colonial Office] wish to avoid the term "permanent inhabitants" in relation to any of the islands in BIOT because to recognise that there are permanent inhabitants will imply that there is a population whose democratic rights will have to

be safeguarded and which will therefore be deemed by the UN Committee of Twenty-Four to come within its purview... It is... of particular importance that the decision taken by the Colonial Office should be that there are no permanent inhabitants in BIOT. First and foremost it is necessary to establish beyond doubt what inhabitants there are at present in the islands, how long they have been resident there and whether any were born on the islands. Subsequently it may be necessary to issue them with documents making it clear that they are "belongers" of Mauritius or the Seychelles and only temporarily resident in BIOT. This device, though rather transparent, would at least give us a defensible position to take up in the Committee of Twenty-Four... It would be highly embarrassing to us if, after giving the Americans to understand that the islands in BIOT would be available to them for defence purposes, we then had to tell them that we proposed to admit that they fell within the purview of the UN Committee of Twenty-Four.[8]

Commenting on a telegram exchange with the UK mission to the United Nations in New York in August 1966, the permanent undersecretary of state for foreign affairs (Sir Paul Gore-Booth, later Baron Gore-Booth of Maltby, GCMG, KCVO) insisted—in a minute marked "confidential"—that

> we must surely be very tough about this. The object of the exercise was to get some rocks which remain ours; there will be no indigenous population except seagulls who have not yet got a committee (the Status of Women Committee does not cover the rights of birds).

This inspired his deputy (Sir Denis Greenhill, later Baron Greenhill of Harrow, GCMG) to add a handwritten note, quipping,

> Unfortunately along with the birds go some few Tarzans or Men Fridays whose origins are obscure, and who are being hopefully wished on to Mauritius etc. When this has been done I agree we must be very tough and a submission is being done accordingly.[9]

The Colonial Office then moved quickly. Within three months after the 1966 UK-U.S. agreement, the governor of the Seychelles (as BIOT commissioner) had two ordinances enacted empowering him to either expropriate or buy out all land for public purposes,[10] and on April 3, 1967, he purchased the Chagos-Agalega coconut plantations for a lump sum of £1,013,200 (about $2.5 million at the time),[11] followed by an arrangement with the managers to lease them back for a windup period during which all services to the islands were gradually reduced to induce the population to leave.[12]

By the end of 1970, when the Pentagon—after some delays due to congressional opposition and protracted rivalries between the U.S. Navy and the U.S. Air Force[13]—finally obtained a $19 million appropriation to start construction in Diego Garcia, there were less than 900 residents left in the archipelago. In a letter to the BIOT commissioner on November 13, 1970 (marked "secret and personal"), the UK Foreign and Commonwealth Office (FCO) indicated that

> we shall continue to try to say as little as possible to avoid embarrassing the United States Administration... We would not wish it to become general knowledge that some of the inhabitants have lived on Diego Garcia for at least two generations and could therefore be regarded as "belongers." We shall therefore advise ministers in handling supplementary [parliamentary] questions... to say that there is only a small number of contract labourers from the Seychelles and Mauritius engaged to work on the copra plantations on the island.[14]

A secret Resettlement Memorandum by the FCO (Exhibit 2), dated January 26, 1971, noted that the U.S. government had confirmed that "their security arrangements at Diego Garcia will require the removal of the entire population of the atoll by July if possible"; as one U.S. State Department official had put it, the island was to be "swept and sanitized."[15]

Exhibit 2   UK Foreign and Commonwealth Office, Resettlement Memorandum, 1971[16]

Secret

From Ian Watt, Atlantic and Indian Ocean Department,	26 January 1971
Foreign and Commonwealth Office, London	[*Received in Registry 16 March 1971*]
To D A Scott [*signature*]/2.2, Sir L Monson, Mr Kerby	HGB 18/2

British Indian Ocean Territory

Resettlement of the Inhabitants of the Chagos Archipelago

Problem

1. The time has come to implement arrangements agreed in principle by the previous Administration by which, in view of the construction of an American naval communication facility on Diego Garcia in the British Indian Ocean Territory (BIOT), we should resettle the population of the Chagos Archipelago, of which Diego Garcia forms part, partly in Seychelles and partly, subject to negotiations with Mauritius Government, in Mauritius.
2. It is also desirable to reaffirm our refusal to issue licences for oil and other mineral exploration within BIOT.

Recommendations

3. I recommend that officials be authorised now to implement the policies described in paragraphs 1 and 2 above; and that we open negotiations with the Mauritius Government about resettlement there, ad referendum to Ministers.

Background and Argument

4. A map of BIOT is at A and a Background Note on Diego Garcia prepared for the British Delegation to the Commonwealth Conference is at B.

Position of the United States

5. The United States Government have recently confirmed that their security arrangements at Diego Garcia will require the removal of the entire population of the atoll by July if possible. This is no surprise. We have known since 1965 that if a defence facility were established we should have to resettle elsewhere the contract copra workers who live there. It is desirable moreover, to arrange for the total evacuation from the Chagos Archipelago of the present population, who are essentially migrant workers. If BIOT is to fulfil the defence purposes for which it was created, there should be no permanent or even

semi-permanent population in respect of which we might in time incur, under Chapter XI of the UN Charter, a variety of obligations including the "sacred trust...to develop self-government."

6. In April 1969 the then Secretary of State obtained the concurrence of the then Prime Minister for the resettlement of the inhabitants of the Chagos Archipelago on the lines now recommended. This was done in the expectation that the US authorities were about to go ahead with their facility. Congress however rejected the proposal, and further action on resettlement was necessarily deferred until American intentions became clear. When the proposal was again submitted to Congress, no approach was made to the Mauritian authorities about resettlement plans in deference to representations by the United States that no overt action should be taken which might prejudice the outcome of Congressional hearings. Now, Congressional approval has been obtained, and the US Government are anxious to begin work.

The Problems of Resettlement

7. There are now about 829 people in the Chagos Archipelago, of whom about 359 live on Diego Garcia itself and the remainder on the two other inhabited atolls of Peros Banhos and Salomon. Of the total, 386 are dual citizens of the United Kingdom and Colonies and of Mauritius (they are known as Îlois). As far as we know, neither the Îlois themselves nor the Mauritius authorities are aware of their dual nationality. There are also 35 citizens of Mauritius, and 408 citizens of the UK and Colonies from Seychelles ("Seychellois"). A population chart is at C.

8. There will be no difficulty in returning the Seychellois to Seychelles, but there would be strong political objections in Seychelles to attempting to settle the Mauritians or Îlois there.

9. In December 1965, a question in the Mauritius Legislative Assembly asked for confirmation that certain obligations in respect of BIOT had been definitely undertaken by the British Government, including, whether all Mauritians then living in Diego Garcia would be resettled in Mauritius; whether the costs of repatriation would be met from the British Exchequer; whether all costs of rehousing them would be met by the British; and finally whether work would be found for them by the British Government. With the approval of the Colonial Office, the Mauritius Government spokesman gave the following answer:

"The British Government has undertaken to meet the full cost of the resettlement of Mauritians at present living in the Chagos Archipelago."

The reply did not refer to the place of resettlement; nor does it ever seem to have been specifically established between the UK and Mauritius Governments that this resettlement would be in Mauritius. But that has been the implication and understanding on our side, and we do not expect the Mauritius Government to dispute it in principle.

10. However, with at least 40,000 men (representing 20% of the labour force) already without work Mauritius has a formidable unemployment problem. In our High Commissioner's view failure by the Mauritius Government to tackle the unemployment could lead to outbreaks of disorder, perhaps comparable to those which in September 1970 led to appeals for British military assistance. The High Commissioner advises that, because of the already high level of unemployment, we must expect negotiations with the Mauritius Government about the resettlement of the people from the Chagos Archipelago to be difficult and the terms demanded high. There are already about 100 families now in Mauritius whose contracts to work in Diego Garcia have not been renewed. The Mauritius Government have been asking how we intend to fulfil our obligations to these people. An answer has been delayed pending a decision about resettlement as a whole.

11. Ideally, we ought to try to settle as many of these people as possible in British territory. There would, however, as stated above, be strong political objections in Seychelles (which has its own economic and over-population problems) to accepting back any other than its own people. The possibilities of resettling Mauritian citizens and Îlois elsewhere in the Indian Ocean area have been re-examined but without success, and it is doubtful if they would wish to go anywhere but Mauritius. However, it may be possible, subject to the concurrence of the Mauritius Government, to resettle as many as 50 Îlois or mono-Mauritian families on the Mauritian island of Agalega, to work on copra plantations run by *Moulinié & Company*, the firm who manage the existing plantations in BIOT. Whatever the result of the proposed negotiations with the Mauritius Government, it will be necessary—if the US deadline for clearing Diego Garcia is to be met—for the inhabitants to be moved temporarily to plantations on Peros Banhos and Salomon, which are also in the Chagos Archipelago. The Governor of Seychelles has confirmed that there will be no practical difficulties in accomplishing this, and that adequate housing and welfare facilities exist for those who are to be moved. Such a measure would, however, be only an interim one.

Costs

12. The total cost of establishing BIOT was originally estimated at £10 million, allocated to the Ministry of Defence Vote. (The US

Government secretly contributed the equivalent of £5 million). Of the £10 million, £3 million was originally allocated for building an airport in Seychelles, £3 million to Mauritius in cash, and £4 million for the purchase of islands from private owners for resettlement and for contingencies. The allocation for the Seychelles airport was notional and in order to build a viable international airport capable of taking modern jet aircraft the cost has risen to over £5 million. (An analysis of expenditure is at D). Consequently, virtually no balance now remains which could be used for resettlement purposes. It is certain that additional funds will be required for resettlement.

13. It is estimated that the cost of resettling the Seychellois in Seychelles may not exceed £10,000 (including costs of termination of contracts and passages). A preliminary figure of £55,000 has been put forward by *Moulinié & Co.* for the proposed scheme for settling 50 families on the Mauritian island of Agalega. The costs of resettlement in Mauritius are at this stage impossible to assess, as we do not know the sort of price the Mauritius Government may ask. The High Commissioner has recommended that a Special Adviser be appointed to examine resettlement possibilities in consultation with the Mauritius Government.

Oil & Minerals

14. Part of the price which the Mauritius Government may ask in any negotiations may be a relaxation of the policy to which we have so far adhered, that no surveying or exploration for oil or other minerals is allowed within the Chagos archipelago, so long as the whole area is set aside for defence purposes. Mauritian interest stems from an assurance given in 1965 that the net benefit of any oil or minerals discovered in or near the Chagos Archipelago would revert to the Mauritius Government. It is not thought that there are any significant resources of oil, gas or minerals in Chagos but some foreign companies have applied for oil exploration licences. It seems right to maintain the ban on mineral and oil exploration for the present, but we should review the need for it with US officials in case it might prove to be in Britain's interest to allow some relaxation in the course of negotiations with Mauritius about resettlement of the Îlois.

15. So far there has been relatively little public or Parliamentary interest in BIOT; and such interest as has been shown has been mainly concerned with conservation on Aldabra (where a nature reserve is being established). But neither we nor the Americans can conceal the fact that the creation of facilities for the US Navy in this British Colony means the evacuation and resettlement of several hundred people;

and if, as seems likely, negotiations with Mauritius are prolonged, the episode may well attract publicity, and critics of our Indian Ocean strategy may be expected to make the most of it.

16. This Submission has the concurrence of Defence Department, Financial Policy and Air Department, Finance Department, East African Department, UN (Policy) Department, Oil Department, the Legal Advisers and of the ODA and the Ministry of Defence. The Treasury have given their concurrence on the understanding that any necessary expenditure over and above the sum of £10 million allocated for the setting up of BIOT will be met from within existing TESC provisions, and subject also to the conditions that resettlement costs shall be kept as low as possible and shall be charged in the first instance to the unspent balance of the sum of £10 million. They also observe that the cost of any further recommendations will have to be considered very carefully, coming as they will on top of the considerable provision already made for establishing BIOT.

[signed: *Ian Watt*]

An advance party of the U.S. Navy (Mobile Construction Battalion 40 and Underwater Demolition Team 12) landed on March 20, 1971, and began construction work on Diego Garcia.[17] On April 16, 1971, the BIOT commissioner issued an Immigration Ordinance that made it unlawful for any person (other than members of the armed forces or public servants) to enter or remain in the Territory without an official permit (Section 9) and provided for the commissioner to make an order directing those persons to "be removed from and remain out of the Territory" (Section 10).

Commenting on an earlier draft of that ordinance, the assistant legal adviser of the UK Foreign and Commonwealth Office (Anthony I. Aust, CMG) had already advised that

> there is nothing wrong in law or in principle to enacting an immigration law which enables the Commissioner to deport inhabitants of BIOT. Even in international law there is no established rule that a citizen has a right to enter or remain in his country of origin/birth/nationality etc. A provision to this effect is contained in Protocol No. 4 to the European Convention on Human Rights but that has not been ratified by us,[18] and thus we do not regard the UK as bound by such a rule. In this respect we are able to make up the rules as we go along and treat the inhabitants of BIOT as not "belonging" to it in any sense.[19]

In an advisory opinion of January 16, 1970, he had explained the purpose of the Immigration Ordinance as "to maintain the fiction that the inhabitants of Chagos are not a permanent or semi-permanent population,"[20] adding,

> The longer that such population remains, and perhaps increases, the greater the risk of our being accused of setting up a minicolony about which we would have to report to the United Nations under Article 73 of the Charter. Therefore strict immigration legislation giving such labourers and their families very restricted rights of residence would bolster our argument that the territory has no indigenous population.[21]

In a minute of January 11, 1971, he further suggested that

> the ordinance would be published in the BIOT Gazette, which has only very limited circulation both here and overseas, after signature by the Commissioner. Publicity will therefore be minimal.[22]

The actual deportation of the remaining 359 "unpeople"[23] from Diego Garcia (comprising 36 Chagossian families and about 100 Seychellian laborers) was organized between July and October 1971 by the British governor of the Seychelles acting as BIOT commissioner, with logistic assistance from the U.S. Navy. On October 15, 1971, a congregation of 35 *Îlois* held their last service in the island church, before being loaded onto the MV *Nordvaer* to be taken to the Seychelles and from there to Mauritius (with one suitcase per person and along with five plantation horses, but without their domestic farm animals).[24] In a bizarre final operation, under personal orders of the commissioner (Sir Bruce Greatbatch, KCVO CMG MBE), the plantation manager and the American soldiers rounded up and exterminated—by rifle, gas, and fire—some 800 stray pet dogs that the islanders had had to leave behind.[25]

In 1975, Congress scheduled a series of hearings on the planned expansion of the Diego Garcia base (which until then had been considered classified, "in response to British sensitivities").[26] Under questioning before the Senate's Armed Services Committee, the Pentagon maintained the agreed fiction that the Chagos islands were "uninhabited" when the U.S. base was established; General George S. Brown, chairman of the joint chiefs of staff, thus testified on June 10, 1975, that Diego Garcia was "an unpopulated speck of land."[27] Only after an investigative report and an editorial published in the *Washington Post*[28] did attention begin to focus on the U.S. role in the expulsion (or, as the *Post* put it, "mass kidnapping") of the Chagossians from their homeland.[29] However, testifying on November 4, 1975, before the Special Subcommittee on Investigations of the House Committee on International Relations, the representative of the U.S. State Department referred to compensation payments for resettlement of the islanders provided by the UK Government to Mauritius in 1973—with funds from the UK-U.S. POLARIS deal—and contended that "these people originally were a British responsibility and are now a Mauritian responsibility."[30]

The United Kingdom had indeed reached an agreement with Mauritius on September 4, 1972, to pay a lump sum of £650,000 ($1.4 million) to cover relief and relocation costs for a total of 1,151 exiled Chagossians. Although payment was made to the Mauritian government in March 1973, funds were not distributed to the islanders until five years later.[31] Compensation awards amounted to about £650 ($1,400) for adults and lesser amounts for children; *no* compensation was ever paid to the other exiled Chagossians in the Seychelles (estimated to number about 500). When the *Washington Post* articles and follow-up stories in *Le Monde* and the London *Sunday Times* revealed that most of the surviving refugees from Diego Garcia were living in abject poverty in the slums of Port Louis,[32] the UK Ministry of Overseas Development commissioned a survey of the Chagossian community in Mauritius in January 1976 (the "Prosser Report"). This report concluded that "the Îlois are living in deplorable conditions"[33] and made a number of proposals for disbursing the £650,000 (the value of which by then had shrunk considerably due to inflation).

Under mounting political pressure and also to forestall the risk of private lawsuits for compensation, the British government in November 1979 offered payment of an additional £1.25 million ($2.7 million) to the Îlois in Mauritius, in exchange for a commitment in writing to "abandon all claims and rights (if any) of whatsoever nature to return to BIOT."[34] Even though abandonment deeds were initially signed by about 1,000 individual claimants,[35] a newly formed Îlois Committee retracted the commitment, and after a series of further negotiations, the UK negotiators agreed instead—on July 7, 1982—to make an additional payment of £4 million ($8.6 million) in "full and final settlement" of all claims against Britain (Exhibit 3, art. 1). The money was to be paid as capital into a trust fund, to which the Mauritian government would contribute land reserved for the Îlois to the value of £1 million; the income of the trust was to be "disbursed expeditiously and solely in promoting the social and economic welfare of the Îlois" (art. 7). Article 4 stipulated that the Mauritian Government, in return, was to procure from the Îlois in Mauritius a signed renunciation of all claims against the United Kingdom for what the agreement delicately called "the events" (as defined in art. 2, i.e., the expulsion).

The Îlois Trust Fund was established by the Mauritian parliament on July 30, 1982; it received the UK government's check on October 22, 1982, followed by a voluntary contribution of one million Mauritian rupees ($180,000) from the government of India. Net payments distributed to 1,344 Îlois by mid-1983—against signature of the required renunciation deeds—amounted to approximately £2,700 each ($5,800) for adults, while some 250 children received about £480 each ($880).[36]

Exhibit 3 UK-Mauritius Îlois Claims Agreement, 1982[37]

Agreement

Between the Government of the United Kingdom of Great Britain and Northern Ireland *And* the Government of Mauritius

concerning the Îlois from the Chagos Archipelago,

signed at Port Louis on July 7, 1982, entered into force on October 26, 1982.

The Government of the United Kingdom of Great Britain and Northern Ireland (hereinafter referred to as the Government of the United Kingdom) and the Government of Mauritius,

—*Desiring to settle* certain problems which have arisen concerning the Îlois who went to Mauritius on their departure or removal from the Chagos Archipelago after November 1965 (hereinafter referred to as "the Îlois");
—*Wishing to assist* with the resettlement of the Îlois in Mauritius as viable members of the community;

—*Noting* that the Government of Mauritius has undertaken to the *Îlois* to vest absolutely in the Board of Trustees established under Article 7 of this Agreement, and within one year from the date of the entry into force of this Agreement, land to the value of £1 million as at 31 March 1982, for the benefit of the *Îlois* and the *Îlois* community in Mauritius;
—*Have agreed as follows*:

Article 1. The Government of the United Kingdom shall *ex gratia* with no admission of liability pay to the Government of Mauritius for and on behalf of the *Îlois* and the *Îlois* community in Mauritius in accordance with Article 7 of this Agreement the sum of £4 million which, taken together with the payment of £650,000 already made to the Government of Mauritius, shall be in full and final settlement of all claims whatsoever of the kind referred to in Article 2 of this Agreement against the Government of the United Kingdom by or on behalf of the *Îlois*.

Article 2. The claims referred to in Article 1 of this Agreement are solely claims by or on behalf of the *Îlois* arising out of:
  (a) All acts, matters and things done by or pursuant to the British Indian Ocean Territory Order 1965, including the closure of the plantations in the Chagos Archipelago, the departure or removal of those living or working there, the termination of their contracts, their transfer to and resettlement in Mauritius and their preclusion from returning to the Chagos Archipelago (hereinafter referred to as "the events"); and
  (b) Any incidents, facts or situations, whether past, present or future, occurring in the course of the events or arising out of the consequences of the events.

Article 3. The reference in Article 1 of this Agreement to claims against the Government of the United Kingdom includes claims against the Crown in right of the United Kingdom and the Crown in right of any British possession, together with claims against the servants, agents and contractors of the Government of the United Kingdom.

Article 4. The Government of Mauritius shall use its best endeavours to procure from each member of the *Îlois* community in Mauritius a signed renunciation of the claims referred to in Article 2 of this Agreement, and shall hold such renunciations of claims at the disposal of the Government of the United Kingdom.

Article 5. (1) Should any claim against the Government of the United Kingdom (or other defendant referred to in Article 3 of this Agreement) be advanced or maintained by or on behalf of any of the *Îlois*, notwithstanding the provisions of Article 1 of this Agreement, the Government of the United Kingdom (or other defendant as aforesaid) shall be indemnified out of the Trust Fund established pursuant to Article 6 of this Agreement against all loss, costs, damages or expenses which the

Government of the United Kingdom (or other defendant as aforesaid) may reasonably incur or be called upon to pay as a result of any such claim. For this purpose the Board of Trustees shall retain the sum of £250,000 in the Trust Fund until 31 December 1985 or until any claim presented before that date is concluded, whichever is the later. If any claim of the kind referred to in this Article is advanced, whether before or after 31 December 1985, and the Trust Fund does not have adequate funds to meet the indemnity provided in this Article, the Government of Mauritius shall, if the claim is successful, indemnify the Government of the United Kingdom as aforesaid.

(2) Notwithstanding the provisions of paragraph (1) of this Article, the Government of the United Kingdom may authorise the Board of Trustees to release all or part of the retained sum of £250,000 before the date specified if the Government of the United Kingdom is satisfied with the adequacy of the renunciation of claims procured pursuant to Article 4 of this Agreement.

Article 6. The sum to be paid to the Government of Mauritius in accordance with the provisions of Article 1 of this Agreement shall immediately upon payment be paid by the Government of Mauritius into a Trust Fund to be established by Act of Parliament as soon as possible by the Government of Mauritius.

Article 7. (1) The Trust Fund referred to in Article 6 of this Agreement shall have the object of ensuring that the payments of capital (namely £4 million), and any income arising from the investment thereof, shall be disbursed expeditiously and solely in promoting the social and economic welfare of the Ílois and the Ílois community in Mauritius, and the Government of Mauritius shall ensure that such capital and income are devoted solely to that purpose.

(2) Full powers of administration and management of the Trust Fund shall be vested in a Board of Trustees, which shall be composed of representatives of the Government of Mauritius and of the Ílois in equal numbers and an independent chairman, the first members of the Board of Trustees to be named in the Act of Parliament. The Board of Trustees shall as soon as possible after the end of each year prepare and submit to the Government of Mauritius an annual report on the operation of the Fund, a copy of which shall immediately be passed by that Government to the Government of the United Kingdom.

Article 8. This Agreement shall enter into force on [the twenty-eighth day after]* the date on which the two Governments have informed each other that the necessary internal procedures, including the enactment of the Act of Parliament and the establishment of the Board of Trustees pursuant to Articles 6 and 7 of this Agreement, have been completed.

---

* Deleted pursuant to an exchange of notes dated October 26, 1982.

In witness whereof the undersigned, duly authorised thereto by their respective Governments, have signed this Agreement.

Done in duplicate in Port Louis this 7th day of July 1982.

For the Government of the United Kingdom of Great Britain and Northern Ireland:	For the Government of Mauritius:
J. Allan	*Jean Claude de l'Estrac*

In 1984, some of the Chagossians also submitted a compensation claim to the U.S. government and brought a class action for damages under the Alien Tort Claims Act against the United States and several of the government officials and contractors involved (including as defendants former secretaries of defense McNamara, Rumsfeld, Laird, and Schlesinger, as well as Halliburton Corp.).[38] The U.S. government first applied to the Mauritian courts for an injunction against the suit—which the Mauritian Supreme Court (Justice Keshoe P. Matadeen) formally refused on August 7, 2002, allowing the plaintiffs to proceed.[39] The U.S. Federal Court of Appeals for the District of Columbia (opinion filed by Judge Janice R. Brown, one of her first cases after appointment by President George W. Bush) ultimately dismissed their claim for damages on April 21, 2006, on the grounds that the establishment of the military base at Diego Garcia was a "non-justiciable political question," adjudication of which would "require the court to judge the validity and wisdom of the Executive's foreign policy decisions," and that the individual defendants enjoyed statutory immunity, as they had been obeying superior orders (*Olivier Bancoult et al. vs. Robert S. McNamara et al.*).[40] The appellate judgment was upheld by the U.S. Supreme Court, denying a rehearing on July 11, 2006, and denying *certiorari* on January 16, 2007.[41]

While the District of Columbia Court of Appeals conceded that the political question doctrine is "murky" and that judicial review might indeed limit the foreign policy and national security powers of the executive branch, it went on to draw a dogmatic distinction between "constitutionally protected liberties" and ordinary statutory rights not so protected[42]—a distinction that is not unchallenged in U.S. constitutional law[43] and that faintly echoes the controversy over "prerogative rights of the crown" invoked by the executive (i.e., claimed to lie outside the scope of judicial review) in the parallel British Bancoult cases then pending.

The islanders—most of whom have since acquired British (and EU) citizenship[44]—had indeed challenged Section 4 of the 1971 BIOT

Immigration Ordinance (as amended in 1981) prohibiting their return to the archipelago. In 1998, Olivier Bancoult (leader of the *Group Refizié Chagos*) brought an action for judicial review of the ordinance in the London High Court (Queen's Bench, Administrative Divisional Court), with a view to enabling the Chagossians to return, and to resettle at least some of the outer islands.[45]

Alarmed by that prospect, the U.S. Department of State—without awaiting the outcome of the lawsuit, which began to make progress—sent a stern warning letter to the UK Foreign Office on June 21, 2000, expressing its

> serious concern over the inevitable compromise to the current and future strategic value of Diego Garcia that would result from any move to settle a permanent resident population on any of the islands of the Chagos archipelago... If a resident population were established on the Chagos archipelago, that could well imperil Diego Garcia's present advantage as a base from which it is possible to conduct sensitive military operations that are important for the security of both our governments but that, for reasons of security, cannot be staged from bases near population centers.[46]

After pointing out the risks of surveillance, monitoring, and electronic jamming as well as from attacks by terrorists operating by boat from the outer Chagos islands of Peros Banhos and Salomon [135 miles = 218 kilometers north of Diego Garcia], the State Department hinted at "a serious re-evaluation of current and future U.S. defense plans involving the island as well as U.S. strategic planning more generally," in the event of "any settlement of a resident civilian population even on the outer islands of the archipelago."

Nevertheless, the London High Court (Lord Justice Laws and Justice Gibbs) on November 3, 2000, quashed the 1971 ordinance on the grounds that the exclusion of an entire population from its homeland lay outside the purposes of the BIOT Order of 1965, which were limited to the governance of the population and did not encompass its expulsion (*The Queen, ex parte Bancoult, vs. Foreign and Commonwealth Office*).[47] Accordingly, a new immigration ordinance was enacted (No. 4 of 2000), which exempted from the need for an entry permit anyone who was a BIOT citizen (i.e., provided the applicant or one of his/her parents or grandparents had been born in the Chagos islands, and including his/her spouse and dependent children). On the other hand, civil proceedings for compensation and for a declaration of the

Îlois' right to return were struck out in a subsequent high court judgment (Justice Ouseley) on October 9, 2003 (*Chagos Islanders vs. Attorney General and BIOT Commissioner*),[48] upheld by the court of appeal in 2004 (Lord Justices Butler-Sloss, Sedley, and Neuberger).[49]

On June 10, 2004, the British government reversed its initial acceptance of the high court judgment of November 2000 and once again intervened to prevent resettlement of the archipelago, by enacting—under its alleged "prerogative" powers, without parliamentary participation—two new Orders-in-Council, the BIOT Constitution Order of 2004 and the BIOT Immigration Order of 2004, which avoided the specific expulsion language of the earlier order but instead formally declared that *no* person had the right of abode in the Territory nor the right without authorization to enter and remain there.[50] As explained in a ministerial statement, the combined effect of the two orders was "to restore full immigration control over all the islands," extending "to all persons, including members of the Chagossian community."[51] After the UK Foreign Office, in response to parliamentary questions in July 2004, categorically denied having received any representations on the issue from the United States,[52] the State Department in November 2004 sent a further letter reiterating its concerns:

> We believe that an attempt to resettle any of the islands of the Chagos Archipelago would severely compromise Diego Garcia's unparalleled security and have a deleterious impact on our military operations, and we appreciate the steps taken by Her Majesty's Government to prevent such resettlement.[53]

In the view of the UK and U.S. governments, therefore, BIOT is to remain a colony inhabited solely by non-permanent residents,[54] including—in addition to the foreign occupants of the Diego Garcia base—a fluctuating population of affluent private yacht owners and charterers (mainly from Australia, Britain, Germany, Switzerland, and the United States). They periodically spend the season between monsoons, that is, March to May, at five exclusive anchorage sites in the Chagos Archipelago, upon authorization by the BIOT administration and against payment of a monthly mooring fee of £100 ($148). The FCO's Notice to All Mariners of New Procedures for Yachts Visiting the British Indian Ocean Territory, of December 8, 2006,[55] thus is in stark contrast to the U.S. State Department's grim scenario of a threat of boat-based terrorism precluding any resettlement of Chagossian fishermen on the very same islands.

Yet, the fiction of the "unpopulated archipelago," staunchly defended by the British Foreign and Commonwealth Office, will inevitably come back to haunt its authors in the very near future. When contesting the 200-mile zone claimed by the United Kingdom for BIOT,[56] the government of the Maldives in its notification dated November 29, 2007, to the UN Secretariat referred to Article 121(3) of the Law of the Sea Convention (UNCLOS), which explicitly rules out the establishment of an exclusive economic zone (EEZ) in the case of "rocks which cannot sustain human habitation or economic life of their own."[57] The two governments have until May 2009 to justify their claims in light of that particular article (notorious among UNCLOS experts as "a perfect recipe for confusion and conflict"),[58] and it will be interesting to see how the Foreign Office lawyers can reconcile the UK's claim to a 200-mile zone around the Chagos Archipelago with their own contention that the archipelago does not have "any permanent population"[59] and with their elaborate feasibility studies purporting to show that resettlement of the islands would not be economically sustainable in the long term ("settlement is not feasible").[60]

Manifestly unimpressed by the Foreign Office thesis and yet another warning letter from the U.S. State Department,[61] the London High Court (Lord Justice Hooper and Justice Cresswell) on May 11, 2006, invalidated those sections of the BIOT Orders that prohibited the return of the native islanders.[62] After the court of appeal affirmed the judgment on May 23, 2007 (Lord Justices Clarke, Waller, and Sedley),[63] the UK government brought an ultimate appeal to the House of Lords' Appellate Committee.[64] Hearings in the House of Lords (before Lord Justices Bingham, Hoffmann, Rodger, Carswell, and Mance) began on June 30, 2008,[65] and the final verdict was delivered on October 22, 2008, accepting—by a 3:2 majority decision—the government's appeal (see Chapter 6 and appendix X). According to the House of Commons' Public Accounts Committee, the consequential *legal costs* incurred by the UK government since 2000 are in excess of £2.171 million so far ($3.21 million),[66] not yet counting the costs of the House of Lords appeal, which is estimated at another £500,000 ($740,000). The legal costs to the U.S. government of the related proceedings in American courts since 2001—right up to the Supreme Court in 2007—have not been made public, but it is safe to assume that they were of a similar magnitude.[67]

Chances are that the native Chagossians and their descendants (whose numbers are beginning to outgrow those of the present occupants of Diego Garcia) will now take their case further, before the European

Court of Human Rights in Strasbourg,[68] which is bound to raise more fundamental issues of British (and by association, American) colonial policy. For over 40 years now, the UK Foreign and Commonwealth Office has pursued the strategy of preserving Diego Garcia as a kind of human rights black hole,[69]

- by systematically excluding BIOT from the territorial scope of most human rights treaties ratified by the United Kingdom—be it Geneva Conventions III and IV,[70] the United Nations Convention against Torture and the European Convention for the Prevention of Torture,[71] or the Statute of the International Criminal Court[72]—and
- by contending more generally that Britain's ratifications of the UN Human Rights Covenants[73] and the European Human Rights Convention[74] do not apply to BIOT because (or so the official reasoning goes) "there is no settled population"—while even openly conceding that "the decisions made by successive governments in the 1960s and 1970s to depopulate the islands do not, to say the least, constitute the finest hour of UK foreign policy" (Bill Rammell, parliamentary under-secretary of state for foreign and commonwealth affairs).[75]

In the view of less charitable commentators, the deportation of the Chagos islanders was "one of the worst violations of human rights perpetrated by the UK in the 20th century" (David R. Snoxell, former British high commissioner to Mauritius and deputy BIOT commissioner),[76] with the complicity of the United States showing "a clear lack of human sensitivity" (Senator Edward M. Kennedy), "oblivious to violations of human rights" (Senator John C. Culver).[77]

CHAPTER 3

Power Politics: "Our" Ocean

Construction of the Diego Garcia military base proceeded in several stages, in response to the ups and downs of global politics, from the cold war to the Middle East conflicts, and from what was first presented as an "austere communications facility"[1] to what was to become one of the most important U.S. military bases or "forward operation sites" overseas.[2] In effect, it turned the Indian Ocean—once known as the "British lake east of Suez"[3]—into an American lake, strikingly reminiscent of the Roman Empire's *Mare Nostrum* (Our Sea).[4]

During the initial phase from 1970 to 1987, total budgetary appropriations allocated by Congress for military construction in Diego Garcia thus amounted to $668.4 million, starting out with one of the largest peacetime projects ever undertaken by the U.S. Navy's Mobile Construction Battalions (CBs = "*Seabees*," at a cost of about $200 million over 11 years)[5] and followed by a series of civilian procurement contracts with U.S. and foreign companies, including the Taiwanese Retired Servicemen Engineering Agency ($6.1 million)[6] and the Japanese Penta-Ocean Construction Co. Ltd. ($17.9 million).[7] The single most important contract, for a total of over $441 million, went to a joint venture of Texas-based Raymond International, Brown & Root (RBRM, a Halliburton subsidiary),[8] and Mowlem & Co. PLC of Middlesex/England, in July 1981.[9] Construction under that contract is well documented, albeit censored for military security reasons;[10] however, the U.S. General Accounting Office reported in May 1984 that unnecessary costs were incurred and construction delays exacerbated because of ineffective management and information and materials control systems.[11] From 1984 on, civilian contractors also took over the maintenance of general facilities and services at Diego Garcia, under a *base*

Table 1 U.S. Congress, Diego Garcia Construction Budget 1970–1987 (in US$ millions)

Fiscal Year	Requested from Congress			Enacted		
	Authorization	Appropriation	Public Law	Authorization	Appropriation	Public Law
1970	9.556 (N)	9.556	91–142	9.556		
1971		5.400 (N)	91–511		5.400	91–544
1972	4.794 (N)	8.950	92–145	4.794	8.950	92–160
1973	6.100 (N)	6.100	92–545	6.100	6.100	92–547
1974S	29.000 (N)	29.000				
1975	3.300 (AF)	3.300	93–552	3.300	3.300	93–636
	14.802 (N)	14.802	93–552	14.802	14.802	93–636
1976	13.800 (N)	13.800	94–107	13.800	13.800	94–138
1978	7.300 (N)	7.300	95–82	7.300	7.300	95–101
1980			96–125	8.600	8.600	
1980S	23.500 (N)	23.500			7.500	96–304
1981	17.977 (N)	17.977				
1981A	104.000 (N)	104.000	96–418	108.177	108.177	96–436
	23.700 (AF)	23.700	96–418	23.700	23.700	96–436
1982	80.130 (N)	80.130				
	78.794 (AF)	78.794				
1982A	122.750 (N)	122.750	97–99	122.750	122.750	97–106
	114.990 (AF)	114.990	97–99	114.990	114.990	97–106
1983	51.639 (N)	51.639	97–321	53.395	53.395	97–323
	36.550 (AF)	36.550	97–321	4.550	4.550	97–323
1984	34.500 (N)	34.500	98–115	31.800	31.800	98–116
	58.200 (AF)	58.200	98–115	58.200	58.200	98–116
1985	6.805 (N)	6.805	98–407	6.310	6.310	98–473
	16.100 (AF)	16.100	98–407	16.100	16.100	98–473
1986	42.680 (N)	42.680	99–167	42.680	42.680	99–173
	5.300 (AF)	5.300	99–167	5.300	5.300	99–173
1987	4.700 (AF)	4.700	99–661	4.700	4.700	

Note: S = Supplement, A = Amendment; N = Navy, AF = Air Force.
Source: Adapted from Bandjunis (Chapter 1, note 6) pp. 309–310, app. 5: "Diego Garcia Military Construction Legislation."

operating service contract,[12] and the monthly shipment of supplies to the base (since 2004 by MV *Baffin Strait* from Singapore).[13] Pursuant to a further exchange of notes in 1987 (appendix VII), all procurement for construction was restricted exclusively to U.S.-UK joint ventures under U.S. management control, with a 20 percent minimum share reserved for UK firms—a clause perhaps best described as a guaranteed kickback arrangement in view of the increasingly lucrative prospects of congressional funding for Diego Garcia.[14] Contractors employed a "third-country" civilian work force of up to 1,900 laborers, mostly from the Philippines, Sri Lanka, and Mauritius (native Chagossians being practically excluded),[15] which at peak periods—together with U.S. military personnel—increased the resident foreign population of the atoll to about 6,000.[16]

As a result, the Diego Garcia coral lagoon has been "cleared" as a turning basin for ships to a depth of 45 feet (13.6 meters),[17] wide enough—and equipped with the necessary wharf and pier facilities—to accommodate an aircraft carrier force of up to 30 ships and nuclear submarines.[18] The airfield has the world's longest slipform-paved runway built on crushed coral (12,000 feet = 3,657 meters, designated as an emergency landing site for the U.S. space shuttle), with ample space to station and service long-distance bombing aircraft,[19] and there are storage facilities for ammunition and supplies to sustain a one-month war effort by a 12,000-troop brigade.[20] A fortuitous British visitor to the atoll in 1984 described it as "a panorama of the American war machine"—a view "that made Pearl Harbor look puny," including lanes of B-52 bombers parked on the airfield and 17 ships riding at anchor in the lagoon, with 13 cargo vessels

> stuffed to the gunwales with tanks and ammunition, fuel and water supplies, rockets and jeeps and armoured personnel carriers, all ready to sail at two hours notice. There was an atomic submarine, the *USS Corpus Christi*...; there was a submarine tender, *USS Proteus* which I had last seen in Holy Loch in Scotland, and which was packed with every last item, from a nut to a nuclear warhead, that a cruising submariner could ever need; and there was the strange white-painted former assault ship, the *USS La Salle*, now converted into a floating headquarters for the US central command, and in the bowels of which admirals and generals played "games of survivable war in the Mid-East theatre," with the white paint keeping their electronic battle directors and intelligence decoders cool in the Indian Ocean sun.[21]

The military metamorphosis of the island did not go unnoticed,[22] and not unopposed. Resistance in Congress focused on the risks of escalation in the global arms race,[23] but was eventually overcome by the Pentagon's dire scenarios of Soviet naval power[24] and by repeated assurances that Diego Garcia was merely to serve as a limited support facility ("not a base").[25] Internationally, the main concern of most littoral states of the Indo-Pacific Ocean—including traditional U.S. allies like Australia[26]—was to prevent the stationing of nuclear weaponry in the region;[27] even Britain's Labour Party passed a motion (at its Blackpool congress in 1982) to "denuclearize" the British Indian Ocean Territory (BIOT).[28] India, Sri Lanka, Mauritius, and the East African coastal countries were adamant on this point, and following a speech in 1970 by Indian prime minister Indira Gandhi explicitly opposing any expansion of the Diego Garcia base,[29] the General Assembly of the United Nations on December 16, 1971, adopted Resolution 2832 (XXVI) declaring the Indian Ocean a "zone of peace."[30] The resolution called on the great powers to enter into immediate consultations with the littoral states of the Indian Ocean with a view to "eliminating from the Indian Ocean all bases, military installations and logistical supply facilities, the disposition of nuclear weapons and weapons of mass destruction," and in 1972, the UN General Assembly established an Ad Hoc Committee on the Indian Ocean to implement the resolution.[31] In 1990, however, France, the United Kingdom, and the United States withdrew from the UN committee, claiming that the cessation of the rivalry of the great powers in the Indian Ocean after the end of the cold war "had rendered the declaration and its purpose obsolete."[32] Since then, the United States has repeatedly called for abolition of the committee for budgetary reasons.[33]

On June 23, 1995, the Organization for African Unity (now the African Union) in turn adopted an African Nuclear-Weapon-Free-Zone Treaty, named "Treaty of Pelindaba"—after the former South African nuclear reactor—and opened for signature at Cairo on April 11, 1996.[34] The treaty requires each party "to prohibit in its territory the stationing of any nuclear explosive devices" while allowing parties to authorize visits or transits by foreign nuclear-armed ships or aircraft (art. 4); furthermore, the three protocols to the treaty require parties not to "contribute to any act which constitutes a violation of this treaty or protocol" (art. 2). According to the map annexed to it, the treaty explicitly covers the "Chagos Archipelago—Diego Garcia," albeit with a footnote (inserted at the request of the United Kingdom) stating that the territory "appears without prejudice to the question of sovereignty."[35]

To date, the Pelindaba Treaty has received 26 of the 28 ratifications required and is expected to come into force in 2009. It was ratified by Mauritius on April 24, 1996; however, both the United Kingdom (which signed both protocols and ratified them on March 19, 2001) and the United States (which cosigned the protocols on April 11, 1996, but so far did not ratify) have since stated their own unilateral interpretation of the footnote to the annex—to the effect that the treaty "does not apply to the activities of the United Kingdom, the United States or any other state not party to the treaty on the Island of Diego Garcia or elsewhere in the British Indian Ocean Territory."[36] Following that peculiar variant of the "black-hole" logic, if Diego Garcia could be excluded from the territorial scope of a treaty by way of a simple reservation on the question of sovereignty, then the 1959 Antarctic Treaty—which also reserves the territorial sovereignty issue—would equally *not* apply to activities in any Antarctic areas where sovereignty is contested.[37] Yet, the U.S. Arms Control and Disarmament Agency considers the footnote in the Pelindaba Treaty map adequate to "protect US interests because any resolution of the [sovereignty] issue will occur outside the framework of the treaty."[38]

Paradoxically, the collapse of Soviet naval power—which had been the primary justification for the Pentagon's establishment and fortification of the Diego Garcia base[39]—was followed by yet another upgrading of the base, in anticipation of new strategic challenges arising from U.S. involvement in the Middle East. Cumulative U.S. investments in the military infrastructure of the atoll have long passed the billion-dollar threshold and continue to rise, with the next five-year upgrading phase estimated at $200 million.[40] This point may be illustrated by two specific items of the arsenal currently stationed there:

1. *Nuclear weapons*: Whereas in the 1980s there were only speculations as to whether the United States deployed nuclear warheads at Diego Garcia,[41] the movement of nuclear-tipped missiles to and from the island by ships or aircraft now seems to be considered "use of the facility in normal circumstances," which under Article 3 of the 1976 UK-U.S. Agreement (appendix IV) no longer requires intergovernmental consultations, but mere advance information from the U.S. commanding officer to the UK officer-in-charge.[42] For example, according to British press reports, over 100 U.S.-made Harpoon missiles were flown to Diego Garcia in September 2003; Israel's three German-built Dolphin submarines then made a port call at the atoll in October 2003 to be fitted with the missiles (which can carry a 200-kilogram nuclear warhead), before returning to the Gulf of Oman.[43] The technical modification of the Dolphin submarines adapting them to the Harpoon missiles had been funded and tested by the German government, which also agreed in 2006 to deliver two more submarines with the same specifications. Doubts remain as to the compatibility of the deal with German obligations under the 1998 European Union Code of Conduct on Arms Exports,[44] and with German

and U.S. obligations under the 1968 Nuclear Non-Proliferation Treaty,[45] the "guidelines for sensitive missile-relevant transfers" (as revised on January 7, 1993) of the 1987 Missile Technology Control Regime, and the 2002 Hague Code of Conduct against Ballistic Missile Proliferation.[46]

On March 30, 2007, the Colorado-based construction firm DGM21 LLC received a further $31.9 million contract for wharf improvements and shore support facilities in Diego Garcia, to accommodate the U.S. Navy's nuclear-powered SSGN attack submarines and a submarine tender.[47] Since the base is not listed among the "inspectable sites" of the 1991 Strategic Arms Reduction Treaty (START-1),[48] forward placement of the Ohio-class special-purpose submarines in the atoll would, in the view of Russian observers, "avoid violating the legal language of START-1 while undermining its spirit."[49]

In May–June 2008, Diego Garcia also served as transit point for 550 tones of low-grade uranium from Iraq, which the U.S. Department of Defense flew out of Baghdad in 37 cargo plane loads for transshipment to Canada by sea (on board the U.S. Navy's crane-ship *SS Gopher State*), in a $70 million operation code-named "McCall."[50] The volume of nuclear material and nuclear weaponry continuously transiting the Diego Garcia base poses considerable risks of radioactive contamination, especially to the waters of the lagoon. Curiously, however, neither the U.S. Navy's 1997 and 2005 Natural Resources Management Plans for Diego Garcia nor the BIOT Conservation Management Plan (2003) make any reference to the need for radiation monitoring.[51] Even though U.S. nuclear-powered submarines are well known to have experienced radiation leakages elsewhere (e.g., in Japan, 2006–2008),[52] the BIOT Administration never took radiation measurements in the Diego Garcia lagoon.[53]

2. *Strategic bombers*: Whereas in 1974 State Department and Pentagon officials had assured Congress that they were "not planning to operate B-52s from or station B-52s at this limited support facility,"[54] the first long-range strategic bombers (B-52 Stratofortresses) arrived in Diego Garcia immediately after completion of the new runway in August 1987.[55] As pointed out later by the State Department, the BIOT base has since played a "central role" in B-52 offensive combat missions to Iraq from 1991 to 1998, during operations Desert Shield, Desert Storm, Desert Thunder, and Desert Fox;[56] and was reportedly used as a staging

area for 20 B-52s deployed as a "calculated-ambiguous" tactical nuclear deterrent against any chemical or biological weapons use by Iraq against U.S. forces.[57] From September 2001 to August 2006, the airfield served as basis for the strategic bombing of Afghanistan—and of Iraq in 2003—by B-52H (Buffs), B-1B (Lancers), and B-2 (Spirits) aircraft. Long-distance bombing sorties from Diego Garcia were discontinued in 2006 when the U.S. Air Force discovered—after five years of operation—that it could save $362,000 per day by stationing the bombers in Qatar and Oman instead.[58] In October 2007, however, the Pentagon submitted a new request to Congress for "supplemental emergency funding" to construct facilities on Diego Garcia for modified B-2 stealth bombers capable of carrying 30,000-pound (13.6-tonne) "massive ordnance penetrator" (MOP) bombs, an upgrade interpreted by British analysts as part of the preparations for potential future intervention in Iran.[59]

Not surprisingly then, much of the contemporary information on Diego Garcia is either classified or "redacted" for reasons of confidentiality.[60] A number of documents reproduced or used in this book became publicly available only after expiry of the 30-year limitation period of the government archives concerned or after formal freedom-of-information requests under the applicable U.S. and UK legislation; a number of other relevant documents known to exist are still restricted.[61] What also became clear in this process is the fact that in several instances the official refusal to disclose "sensitive" information was based less on legitimate concerns of national security but rather on a pervasive desire of government authorities to shield their politicians, civil servants, and military staff—at best, from embarrassment for past errors of judgment, and at worst, from accountability for past transgressions of national or international law.

CHAPTER 4

Military Secrecy: Public Access Denied

As confirmed by the U.S. State Department's admonitions to the UK Foreign Office in 2000 and again in 2004,[1] American objections to any resettlement of the British Indian Ocean Territory (BIOT) were in large part motivated by security concerns. This was in view of the more or less secretive communications and intelligence projects operating at Diego Garcia under U.S. Air Force, Navy, and CIA auspices, variously code-named "Kathy," "Reindeer," "Charlie," et cetera.[2] One of the key assets of the island—"lonely bastion in the vastness of the Indian Ocean"[3]—indeed was its very remoteness and isolation, shielding the base from unwanted observation and public attention. A Foreign Office memorandum to Lord Douglas Hurd, Secretary of State for Foreign and Commonwealth Affairs, recommended that "no journalists should be allowed to visit Diego Garcia" and that visits by parliamentarians or congressmen be kept to an absolute minimum to keep out those "who deliberately stir up unwelcome questions."[4] Not surprisingly perhaps, in response to questions about the expulsion of the Chagossian islanders and any political or media scrutiny occurring while the Diego Garcia construction project was proceeding, a U.S. Navy spokesman declared in 1988,

> On the matter of public debate, nobody seemed concerned. There was not much media involvement. The U.S. Navy were too busy building it. It was satisfying to get a little recognition from time to time, but there was no controversy over the program.[5]

The first operational project established on the atoll in 1971 was a joint UK-U.S. signals and intelligence (SIGINT) station to monitor naval radio traffic in the Indian Ocean,[6] as part of the global surveillance network (later code-named "Echelon") formed by the five Anglo-Saxon intelligence-collection agencies—Britain's Government Communications Headquarters (GCHQ), the United States' National Security Agency (NSA), Canada's Communications Security Establishment (CSE), Australia's Defence Signals Directorate (DSD), and New Zealand's Government Communications Security Bureau (GCSB)—cooperating since 1947 under the so-called (still unpublished) UKUSA Agreement.[7] In the spring of 1973, when the NSA anticipated that it would have to abandon its Kagnew intercept station in Asmara/Eritrea for political reasons,[8] it sent an exploratory mission (code-named "Jibstay") to Diego Garcia, with a team of 14 intercept operators and analysts from military cryptologic organizations, to conduct "hearability tests" and to set up an intercept antenna "farm" on the southern tip of the island.[9] A direction-finding (DF) signals interception station, part of a worldwide network code-named "Classic Bull's-Eye," began operating from Diego Garcia in August 1974.[10] From 1976 on, the station also served as a ground control base for the UKUSA Advanced Tactical Ocean Surveillance System (code-named "Classic Wizard"), part of the U.S. Navy's highly secret Naval Ocean Surveillance System program (NOSS, "White Cloud"), using satellites and low-to-medium frequency underwater signalling (IUSS, Integral Undersea Surveillance System) to monitor ships and submarines worldwide[11] and for covert monitoring of electronic emissions under a Galactic Radiation Background (GRAB) satellite experiment.[12]

> Secrecy procedures at the Ocean Surveillance building in Diego Garcia did not prevent one of the naval radio technicians there (who served as supervisor of technical control in 1974–1975) from gaining access to classified messages and cryptograph lists and selling them to the Soviet KGB. It enabled the Russians, with the help of electronic hardware obtained during the 1968 *Pueblo* incident,[13] to read all cryptographic traffic between U.S. Naval Headquarters and ships around the world—a scoop later ranked by a top KGB veteran as "the greatest achievement of Soviet intelligence at the time of the Cold War."[14] The navy spy was arrested in 1985 and sentenced to a life-time prison term, affirmed on appeal in 1987.[15]

Starting in 1987, the U.S. Air Force operated a site of its Ground-Based Electro-Optical Deep Surveillance System (GEODSS) on the atoll[16] for the tracking of man-made space objects—somewhat belatedly approved by the UK Foreign and Commonwealth Office in December 2004 (see appendix IX/3).

According to the 1966 "Agreed Confidential Minutes" [appendix II, para. II], *prior approval* was explicitly required for installation of a "space tracking station"—which seems to have escaped the attention of the UK authorities for over seven years.

The Diego Garcia antenna farm also serves as one of the five ground control bases assisting in the operation of the Global Positioning System (GPS, officially "NAVSTAR-GPS") managed by the U.S. Air Force, operational since 1995. Following a further UK-U.S. exchange of notes in July 1999 (appendix VIII), the hydro-acoustic monitoring station installed on Diego Garcia/BIOT became part of the International Monitoring System established under the Comprehensive Nuclear Test Ban Treaty.[17] Since February 2004, Diego Garcia (coded "DGAR") has also served as one of 40 stations within the International Deployment of Accelerometers system of the IRIS Global Seismographic Network, installed by civilian contractors. Its technical equipment consists of seismometers, a strong-motion sensor, and a microbarograph—adjacent to the air force's GEODSS installation—to record any shaking caused as seismic waves travel through the earth.[18]

On December 26, 2004, DGAR was among the first stations to pick up the catastrophic Andaman earthquake that triggered the "Boxing-Day *tsunami*," resulting in more than 225,000 deaths in 11 coastal countries of the Indian Ocean (including at least 149 British and 33 American tourists). Diego Garcia was narrowly spared from disaster and merely recorded a 6-foot (1.8 meters) tidal surge with minor damage on the eastern rim, because the special topographic conditions of the atoll prevented the *tsunami* from building up before passing the island and moving on toward East Africa.[19] It has been alleged that DGAR, as one of the three IRIS stations closest to the quake's epicenter (see figure 1), could have alerted the regions affected [and the thousands of U.S. servicemen and foreign workforce then stationed at the base] in time and failed to do so; however, most subsequent analyses attribute the tragic failure of all existing warning networks in that case to systemic defects in automated monitoring and risk classification for exceptional events of this magnitude.[20]

A major unresolved security issue is the role of the Diego Garcia base—or of U.S. naval vessels anchored in the lagoon—as a prison site

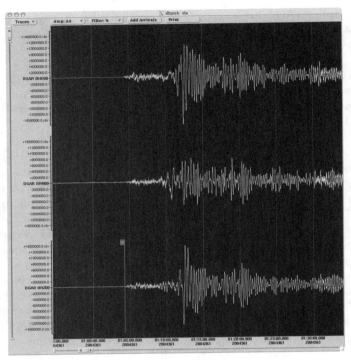

Figure 1 IRIS/IDA Seismic Data Diego Garcia, *Tsunami*, 2004[21]

for suspected foreign terrorists. Detention of such prisoners on UK territory would be incompatible with Britain's obligations under the 1984 UN Convention against Torture (in particular, art. 3 on "non-*refoulement*")—which, however, does not seem to have been formally extended to BIOT so far.[22] According to a report of the U.S. General Accounting Office, construction works on Diego Garcia included a "detention facility" completed as early as 1983,[23] built under contract by the same firm (Halliburton subsidiary Kellogg, Brown & Root) that in 2005 received a $30 million award for construction of the infamous Detention Camp #6 at U.S. Naval Base Guantánamo.[24] On June 21, 2004, UK foreign secretary Jack Straw stated—in response to parliamentary questions—that the BIOT commissioner "has declared certain specified premises in Diego Garcia to be a prison" by orders made in February 1986 (replacing an order of July 1982), July 1993, and December 2001.[25] However, the parliamentary under-secretary of state for foreign and commonwealth affairs specified on July 20, 2004, that "the only United States prison facilities on Diego Garcia are those established under U.S. service law for the detention, as authorized by that law, of members of the United States armed forces and any accompanying civilian personnel and dependents."[26]

Following public statements by retired U.S. General Barry McCaffrey to the effect that suspected terrorists *were* being held at Diego Garcia[27] and allegations in the Council of Europe's Parliamentary Assembly that the base had been among the destinations of secret "extraordinary rendition" flights by U.S. aircraft,[28] several UK government officials issued categorical denials:

> I repeat that the stories which have appeared in the press are completely without foundation. The United States Government would need to ask our permission to bring suspects to Diego Garcia, and they have not done so. No suspected terrorists are being held on Diego Garcia and, under current BIOT law, there would be no authority for the detention of *Al'Qaeda* suspects in the territory. (January 8, 2003)[29]

> The US authorities have repeatedly assured us that no detainees have at any time passed in transit through Diego Garcia or its territorial waters or have disembarked there and that the allegations to that effect are totally without foundation. The Government are satisfied that their assurances are correct. (June 21, 2004)[30]

> The US authorities have repeatedly given us verbal assurances, most recently in May 2004, that no prisoners have at any time passed in transit through Diego Garcia or its territorial waters or have disembarked there. The British Representative on Diego Garcia has confirmed this to be the case. (November 2, 2004)[31]

> There is no US facility for foreign detainees on Diego Garcia. The only civilian detention centre is at the small UK-run police station. The US authorities have repeatedly given us assurances that no detainees, prisoners of war or any other persons in this category are being held on Diego Garcia or have at any time passed in transit through Diego Garcia or its territorial waters or airspace. This was most recently confirmed during the 2006 UK/US political-military talks held in London on 17 and 18 October. (October 26, 2006, reiterated *verbatim* on October 11, 2007)[32]

> The US authorities have repeatedly given us assurances that no terrorist suspects have been, or are being held at Diego Garcia, or at any time have passed in transit through Diego Garcia or its territorial waters or airspace. (July 18, 2007)[33]

However, after a phone call from Secretary of State Condoleezza Rice, Foreign Secretary David Miliband conceded on February 21, 2008, that Diego Garcia had indeed been identified as the destination of at least two U.S. "rendition flights" in 2002—for purposes of "refueling," according to CIA director (U.S. Air Force General) Michael Hayden.[34] (It is of course plausible that after taking prisoners—or "detainees," in CIA terminology—5,000 miles [8,000 km] to the middle of the Indian Ocean, a Gulfstream jet landing in Diego Garcia would have needed refueling).[35]

Answering questions in the House of Commons as to whether any suspected terrorists were held on U.S. "prison hulks" at or near Diego Garcia,[36] Margaret Munn (parliamentary under-secretary of state for foreign and commonwealth affairs) firmly—if cryptically—responded,

> We have no evidence to suggest that any ships outside the territorial waters of Diego Garcia have been involved in rendition, nor that they have been serviced from the island...There is an issue about how far outside the territorial waters they are if they are outside them...If they are outside our territorial waters, our responsibilities are different from if they are inside. (March 26, 2008)[37]

Additional information confirming U.S. detentions and "interrogations" on or off Diego Garcia between 2003 and 2006 was made public on July 31, 2008.[38] However, efforts to obtain further clarification and documentation have been unsuccessful so far. In particular, the UK Foreign Office has refused to release to British parliamentarians the secret minutes of the annual UK/U.S. political-military talks on Diego Garcia,[39] on the grounds that doing so would jeopardize national security and might damage British-American diplomatic relations.[40]

Sadly, the tragic events of September 11, 2001, seem to have dealt a major blow to the principle of public access to environmental information, too. The United States, once the undisputed champion of open government and transparency in this field,[41] has begun to sound the retreat from a citizens' right-to-know toward a restrictive "need-to-know" approach for security reasons[42] and has effectively withdrawn from most international efforts at environmental information disclosure.[43] And even though the United Kingdom is a party to the 1998 Aarhus Convention on Access to Information, Public Participation in Decision-Making and Access to Justice in Environmental Matters (still boycotted by the U.S. State Department),[44] the Foreign and Commonwealth Office takes the position that the convention "has no practical relevance to BIOT" because "BIOT has no permanent residents."[45] As a result, both the United States and Britain have consistently followed a "hiding-hand" strategy also with regard to major environmental issues concerning Diego Garcia.

The problem is exemplified by the "undersea noise pollution" issue. In the context of secret programs going back to cold war anti-submarine strategies, the U.S. Navy has long conducted low-frequency underwater sound monitoring operations and experiments at Diego Garcia.[46] The Diego Garcia hydrophone station (HA-08) also participated in experimental programs in the Southern Indian Ocean in 2001 and 2003, in the course of which underwater explosive charges were test-fired in order to measure long-range low-frequency sound propagation.[47] The potential harmful effects of military low-frequency active sonar (LFAS) on marine life—especially cetaceans—have been known and discussed since the 1980s,[48] highlighted by a series of mass strandings of Cuvier's beaked whales (*Ziphius cavirostris*) in the Canary Islands (1985, 2002), Japan (1988–2004), Greece (1996), the Bahamas (2000) and California (2002)—invariably in connection with underwater active sonar operations at or near U.S. naval bases or by NATO vessels.[49] The Chagos Archipelago boasts its own population of a sub-species of Cuvier's beaked whale (*Ziphius cavirostris indicus*),[50] as well as breeding grounds for sperm whales (*Physeter macrocephalus*, protected on Appendix I of the 1979 Migratory Species Convention)[51] and other cetaceans.[52] The entire archipelago is part of the "Indian Ocean Sanctuary" declared under the International Whaling Convention in 1979.[53] Yet, owing to the military secrecy shrouding the Diego Garcia "ocean surveillance station" (situated immediately adjacent to the "restricted conservation area" established by BIOT Ordinance No. 6 of 1994, in force November 24, 1997), there has been no assessment so far—whether by American or by British authorities—of the station's actual or potential impacts on cetaceans in BIOT.[54]

CHAPTER 5

Nemesis: Natural Heritage Dredged—and Drowned

From an environmental perspective, Diego Garcia's "chief natural glory"[1] is its unique coral reef—with well over 100 species of coral, among the richest stores of biodiversity in the Indian Ocean.[2] In 1994, the UK government declared the eastern part of the reef (including three islets at the entrance of the lagoon) a "restricted conservation area," that is, entry was forbidden except as expressly authorized by permit.[3] The BIOT Conservation Policy Statement issued in October 1997 specified that the entire archipelago was to be managed in accordance with the standards of the 1972 World Heritage Convention,[4] that is, "the islands will be treated with no less strict regard for natural heritage considerations than places actually nominated as World Heritage sites, subject only to defence requirements."[5] And by 1999, the UK Joint Nature Conservation Committee boldly claimed that "because of its military status, the whole of BIOT acts as a *de facto* protected area."[6] Also because of the U.S. military facility, the UK Foreign and Commonwealth Office (FCO), until the 1980s at least, had tried to suppress "any mention of Chagos in scientific reports"[7] and in 1994 successfully vetoed the inclusion of BIOT in Britain's ratification of the 1992 Convention on Biological Diversity.[8] Over considerable opposition from the FCO, the 1971 Ramsar Convention on Wetlands of International Importance (ratified by the UK in 1976)[9] was extended to BIOT more than 22 years later, by declaration on September 8, 1998. It took the FCO another three years (until July 4, 2001) to register the Diego Garcia "conservation area" as an internationally protected wetland on the convention list (including the lagoon and territorial waters of the island)—though expressly *not including* "the area set aside for military uses as a U.S. naval support facility";[10] i.e., creating a new legal black-hole, by carving out the land area of the U.S. base and the adjoining western reef flats—as delimited in the official Ramsar map (see Map 3)—from the scope of the treaty. Nonetheless, naval operations in the lagoon and the atoll's three-mile territorial sea have since 2001 become subject to the rules and restrictions of the Ramsar Convention (to which the United States is a party, too).

The construction work undertaken in the military base over the past 37 years inevitably left its mark on the atoll. To begin with, several thousands of trees were bulldozed down.

A report by a visiting navy natural resources specialist in December 1973 lamented the "eyesore" created by the unusable remnants of "a thousand or more coconut trees" left over from the first clear-cut.[11] When another 1,000-plus trees were removed in 2006–2007 to make way for a new ammunition pad, the now-resident natural resources specialist for Diego Garcia received the *FY07 Chief of Naval Operations Environmental Award* for his efforts to relocate 265 threatened coconut crabs displaced by the project.[12]

The U.S. Navy's "Seabees" engineers started out in 1971 by dynamiting a larger access channel through the northern reef of Diego Garcia, deep-dredging part of the lagoon and locating sites for "coral harvesting," that is, mining the atoll's coral as aggregate material for harbor and airfield construction and as landfill for the draining and filling of large swamp/wetland areas.[13] One of the former BIOT commissioners has portrayed the 1970s as "the heroic pioneer days of the new military society on Diego Garcia,...a certain 'Wild West' atmosphere."[14] The Seabees (over 10,000 of whom served on the base, with 600-man battalions rotated in nine-month intervals until July 1982)[15] and their diving contractors (Oceaneering International, from Santa Barbara/California) had a reputation for somewhat unorthodox operational practices; for example, surplus unstable dynamite was occasionally disposed of by simply packing and detonating it underwater in the previously blasted holes in the coral reef—with explosions that "rocked the island."[16]

Coral-blasting and dredging operations—in an area of 11.9 square miles (30.8 square kilometers) in the lagoon and in strips along 4 miles (6.4 kilometers) of seaward reef flat—were continued over the next 11 years by the Seabees and under contracts awarded to Taiwanese and Japanese firms specialized in harbor dredging (for a total of $24 million).[17] In 1981, the U.S. Navy awarded a further contract to an Anglo-American joint venture for major expansions of the runway, wharf, and piers and for a carrier-force turning-basin and anchorage in the lagoon.[18] The funding allocated in this context specifically included "dredging," estimated at $13 million for navy projects and $8 million for air force projects in fiscal years 1981–1982.[19] When it turned out that further coral mining alone could not meet aggregate requirements without causing irreparable damage to the reef, limestone aggregate with crushed sand and 157,000 tonnes of cement were imported by sea from Singapore/Malaysia and East Africa instead.[20] Coral dredging continued, though, both in the lagoon (as the navy elegantly put it in 2005, "much of its coral was relocated as part of the dredging effort")[21] and in the western reef flats: dredged "coral fill" available as raw material on Diego Garcia was estimated at 5 million cubic yards (4.5 million cubic meters) by 1983.[22] Overall, an area of about 100 acres (40 hectares) was land-filled, and a total of more than 150,000 cubic yards (115,000 cubic meters) of concrete were poured for the construction of airport runways and parking aprons, 18 miles (29 kilometers) of asphalt road ("Interstate DG-1"), antenna fields, and support facilities.[23] As a result of this Herculean development project, the "downtown area" of the base today is more reminiscent of the Florida Keys than of the Indian Ocean, with all the facilities of a small American town.[24]

It was clear from the outset that this radical transformation of the island would have major, and possibly irreversible, environmental effects. The 1972 Supplement to the 1966 UK-U.S. Agreement on Diego Garcia (appendix III) therefore introduced a "conservation clause" stating that "as far as possible the activities of the facility and its personnel shall not interfere with the flora and fauna of Diego Garcia."[25] A Natural Resources Conservation Land Management Plan drawn up in 1973 by a U.S. Navy biologist recommended a number of measures for nature protection, landscaping, and waste disposal in the American sector of the base,[26] later followed by similar recommendations addressed to construction contractors, "to protect specific resources on Diego Garcia."[27] Yet, it was not until July 1980—that is, more than nine years after the U.S. Seabees started their massive reef-blasting and coral-mining operations—that the navy commissioned a first comprehensive "Environmental Survey of Construction and Dredging Related Activities on Diego Garcia."[28] Except for recommending alternative dredging technologies, alternative disposal sites for "harvested" coral material, and development of a monitoring program,[29] the report in essence concluded that there had been no significant adverse effects and—with regard to future projects planned—that

> if the dredging activity is of the same magnitude as conducted during the operations completed in 1976, then recovery of these coral reefs and associated fish fauna is expected. Nevertheless, care must be taken to conduct dredging activities in a manner that would minimize any impacts.[30]

According to British marine biologists, too, a brief visit in the late 1970s "showed little damage outside the immediate port area despite dredging" in the lagoon.[31] However, they noted the absence of any published observations from the greater port construction work in subsequent years and categorically rejected any predictions of "natural recovery" in the dredged reef flats.[32]

When the next UK–U.S. Supplementary Agreement was signed in London in 1976 (appendix IV), it became clear that the U.S. Navy planned much more radical coral-mining works in the lagoon than ever before—with an area of 4,000 acres (16 square kilometers) now designated for "expanded dredging for fleet anchorage"[33] and with multi-million-dollar dredging contracts awarded from 1981 on to several foreign firms and to the Anglo-American RBRM joint venture in particular.[34] Yet it was only six years later that the BIOT administration seems to have realized what was going on and eventually felt prompted to insert a new paragraph in the 1982 Supplement to the agreement (appendix VI), reading,

> *Dredging and Reef-Blasting.* It will be the responsibility of the U.S. Commanding Officer to ensure that no dredging or reef blasting is conducted in any area where it could cause irremediable damage. Normally, a breadth of live reef of no less than 80 yards [73 meters] will be left intact and free of blasting operations all around the island. If, exceptionally, it is considered necessary to reduce the breadth of the live reef to less than 80 yards at any point, the British Government Representative will be consulted in sufficient time in advance of the proposed action to enable an assessment of the likely ecological consequences to be made by qualified authorities.[35]

The shared responsibilities so invoked inevitably raise the question of the applicable environmental legislation. In 1993, the British conservation adviser to the BIOT administration was intrigued to learn that, in what sounds like the U.S. Navy's own "legal black-hole" theory,[36] "being located overseas, the US Environmental Protection Agency (EPA) regulations do not apply to Diego Garcia."[37] Yet, pursuant to the 1979 U.S. Presidential Order on Environmental Effects Abroad of Major Federal Actions,[38] all federal agencies of the United States are required to undertake environmental impact statements, reviews, or studies for any project likely to cause significant harm to the environment of a foreign nation.[39] That requirement—considered mandatory for "large dredging projects" and "major construction and filling in tidelands/ wetlands," in particular—is also part of the U.S. Navy's Environmental and Natural Resources Program Manual (OPNAVINST 5090.1B),[40] as confirmed in the 1997 Natural Resources Management Plan Diego Garcia (NRMP)[41] and related Final Governing Standards.[42] The commanding officer of the navy support facility at Diego Garcia has indeed acknowledged the duty to consider such reviews and studies in current project preparation practice in Diego Garcia.[43]

It is most surprising, therefore, that the competent U.S. Naval Facilities Engineering Command Pacific (Pearl Harbor) apparently did *not* prepare formal environmental impact statements, reviews, or studies prior to the multi-million-dollar dredging and land-fill projects undertaken in Diego Garcia,[44] which in retrospect were found (by the navy's own natural resources specialists) to have had "a potential to cause deleterious, long-term impacts, including increased beach and shoreline erosion and...adverse impacts on marine life."[45] One confirmed side-effect *not* assessed beforehand was the introduction of invasive alien plant species—including tangan-tangan (*Leucaena leucocephala*, considered one of the top 100 worst invasive species of the world)— with the massive imports of landfill aggregate from Malaysia and by seeds introduced with navy equipment, that is, an ecological "time bomb that should be defused," according to the advice of a navy consultant in retrospect.[46] A 1996 U.S. scientific expedition commented that the past "coral-harvesting" from the seaward side of the reef was "something of an ongoing experiment, with no indication of success or failure in terms of recovery, or protection of the shoreline," whereupon the navy's 1997 Natural Resources Management Plan finally recommended that "no new dredging be authorized without first having careful investigations."[47] The plan further recommended annual follow-up monitoring of the dredged areas after four years—which, however, does *not* seem to have been done to date because "funding has not been identified" for that purpose.[48]

Another serious problem has been chronic pollution by fuel spills from the petroleum, oil, and lubricants (POL) storage tanks on Diego Garcia, which hold 1.34 million barrels (177,000 tonnes) of jet fuel and diesel oil.[49] The first major spill of approximately 1 million gallons (3,000 tonnes) of *JP-5* diesel fuel occurred in the 1980s, possibly as a result of pipeline fractures caused by a 7.6 earthquake in November 1983. By the time the underground leakage was discovered, it had filled and replaced the entire freshwater lens below the airfield,[50] so the leaked fuel could be extracted by scavenging wells drilled near the tanks and reused after re-refining.[51] Three more fuel spills were identified in 1991 (160,000 gallons lost = 685,000 liters), 1997, and 1998;[52] to remedy those spills, "bioslurper" systems were installed, which from March 1996 to October 1997 recovered an estimated 70,000 gallons (300,000 liters) for direct reuse in the diesel-powered electric generating plant of the base, while the rest was considered to have been aerobically degraded in the soil.[53] In terms of magnitude, the spills far exceeded those reported from other U.S. military bases in Panama, Puerto Rico, and the Philippines.[54]

Even though recent water quality measurements in the lagoon show relatively low pollution levels,[55] the non-governmental British *Chagos Conservation Trust* noted in 2004 that "the US Air Force has still not cleared up its oil spills, and the US Navy has been slow in implementing its own *Natural Resources Management Plan* for Diego Garcia,"[56] partly because remediation had to be interrupted in light of new operational requirements. In 2004, the Diego Garcia Naval Support Facility thus earned the American Petroleum Institute's Award for Excellence in Fuels Management, for issuing 88.8 million gallons (266,400 tonnes) of *JP-5* fuel during operations Enduring Freedom and Iraqi Freedom.[57]

An earlier report by the then BIOT conservation adviser (and former British government representative on the base) had concluded in 1996 that

> during all this time, there has been no significant contribution from the USA, who of course have caused significant ecosystem disturbance in developing Diego Garcia. The UK has even undertaken some NRMP items which should have been funded by the USA... The USA is not pulling its weight... Conservation is about the only field of endeavour in which we can earn credit for being in the Indian Ocean where other countries do not want us.[58]

To which the BIOT administration's comprehensive Chagos Conservation Management Plan laconically added in 2003: "This has not noticeably changed in the last six years."[59]

In fairness, though, the state of the environment of Diego Garcia is not solely an American responsibility. The U.S. Navy's Integrated Natural Resources Management Plan of 2005 unequivocally says,

> US federal policies and programs apply only to the extent that the UK agrees that they should be applicable and as they conform to British Indian Ocean Territory (BIOT) policies and programs. The full governmental and civilian judicial authority, including that relating to natural resources conservation and environmental protection, rests with the British Representative (BRITREP), a Senior Royal Navy Commander. The UK, through the BRITREP, generally monitors environmental matters. Larger environmental concerns are referred to the annual US/UK Political Military (Pol-Mil) Talks for resolution.[60]

Under these circumstances, it is all the more difficult to understand the wholly passive role of the UK Foreign and Commonwealth Office, reflected in a long series of public statements—defensive, evasive, or just looking the other way:

> The BIOT Government believes that the American authorities in Diego Garcia are acting in an environmentally responsible manner. (April 26, 1999)[61]

> The United States have produced a Natural Resources Management Plan for Diego Garcia, and we published the Chagos Conservation Management Plan in October 2003. Environmental issues are regularly discussed by the UK and US Governments. (July 12, 2004)[62]

> The BIOT Government and the US authorities collaborate on all aspects of the conservation and protection of the natural environment of Diego Garcia and where appropriate of the outer islands of the Territory. In this context, the US authorities on Diego Garcia have established a Natural Resources Management Plan and they expend considerable resources on environmental conservation. The UK Government attaches considerable importance to the conservation of the natural environment of the Chagos islands and the BIOT Government have ensured that the necessary legislation

for this purpose is in place and is enforced, and that relevant international conventions are observed. (July 19, 2004)[63]

It is worth recalling here that the site of all fuel spills is immediately adjacent to the lagoon listed since 2001 by the UK government as an internationally protected site—excluding the U.S. military facility—under the 1971 Ramsar Convention on Wetlands of International Importance (to which Mauritius and the United States are also parties).[64] According to Article 3(2) of the convention, each contracting party is under an obligation to inform the convention bureau, without delay, "if the ecological character of any wetland in its territory and included in the List has changed, is changing or is likely to change as a result of technological developments, pollution or other human interference." Yet the 2008 UK Information Sheet on Ramsar Wetlands for Diego Garcia, listed as site 1077 (*2UK001*) of the convention, merely indicates under "factors (past, present, or potential) adversely affecting the site's ecological character, including changes in land (including water) use and development projects": n/a = "not applicable because no factors have been reported."[65] That is, *no* reports of any fuel spills, other "adverse factors," ecological changes, or risks (such as dredging, radiation, sonar noise, or power plant construction) to the Diego Garcia wetland area or lagoon have ever been communicated to the bureau of the convention in Gland/Switzerland—or to any other "competent international organizations" (under UNCLOS Articles 204–206) for that matter.[66]

The situation of Diego Garcia is different—albeit worse—under the 1992 Convention on Biological Diversity,[67] which also requires periodic reporting (art. 26) and which requires all member states, inter alia, to "prevent the introduction of, control or eradicate those alien species which threaten ecosystems, habitats or species" (art. 8/h). Yet, since the Foreign and Commonwealth Office has persistently vetoed (because of the U.S. military facility) an extension of that convention to BIOT,[68] the dramatic increase of invasive plant species introduced by navy operations and construction in Diego Garcia (pp. 53–55 above) has never been mentioned in any UK national report.[69] Commenting on the FCO's veto, renowned biologist David Bellamy OBE bluntly observed that "the British Government's attitude is foolhardy in the extreme and shows as usual a total lack of understanding of their international commitments to Ramsar and Rio."[70] As a result, the only parts of the world where the Biodiversity Convention (with 190 member countries the most widely accepted global environmental treaty) is *not* applicable today are the United States, Iraq, Somalia, and four UK overseas territories: Bermuda, the British Antarctic Territory, the Pitcairn Islands, and BIOT.[71] If there ever was a "black hole" in international environmental law, this is it.

Meanwhile, alas, far more serious environmental risks threaten the atoll as a result of global climate change. British scientists have documented an alarming rise of the annual average sea temperature around the Chagos islands during 1990–2006.[72] "Temperature shock" reportedly was the primary cause of severe coral mortality ("bleaching," in the wake of *El Niño*) in 1998, which—together with ocean acidification due to increased uptake of carbon dioxide from the atmosphere—is forecast to lead to even more radical live coral loss in the area in the foreseeable future, to the point where "coral reefs will be the first major ecosystem to be functionally extinguished because of climate change."[73] Measurements also show an accelerating rise of sea levels in the Chagos Archipelago, with concomitant signs of beach erosion damage,[74] in line with current estimates of global sea-level increase ranging from 2.6 to 6.6 feet (0.8–2.0 meters) for the year 2100,[75] which are similar to contemporary observations in other Indian Ocean islands[76] and which in the UK government's "resettlement feasibility study" of 2002 were actually cited among the reasons against allowing the Chagossian islanders to return to their archipelago.[77]

In a recent Washington think-tank report on "National Security and the Threat of Climate Change" (CNA 2007), prepared by a "blue ribbon" military advisory panel of 11 retired U.S. generals and admirals,[78] Diego Garcia was singled out—because of its low average elevation of 4 feet (1.3 meters) above sea level—as the prime example of a "losing place,"[79] urgently "requiring advance military planning...in view of possible climate change impacts over the next 30 to 40 years."[80] Translated from West Point jargon, this presumably means that the base should now be reconsidered for inclusion in the Pentagon's "Global Posture Review,"[81] that is, for terminal closure—at the very moment when the U.S. government has initiated another five-year $200 million construction program to upgrade Diego Garcia for use by the Air Force's modified B-2 stealth bombers as well as by the Navy's new nuclear-powered attack submarines and a submarine tender inside the lagoon (Map 3),[83] with a new power plant using ocean thermal energy conversion (OTEC) to be built on an off-shore platform in the territorial waters of the island,[84] and with U.S. military strategists now advocating an "expanded presence on Diego Garcia"[85] (in other words, yet another surge).

The CNA Panel also valiantly recommended that "the US should commit to a stronger national and international role to help stabilize climate changes at levels that will avoid significant disruption to global security and stability."[86] It seems ironic indeed that the United States

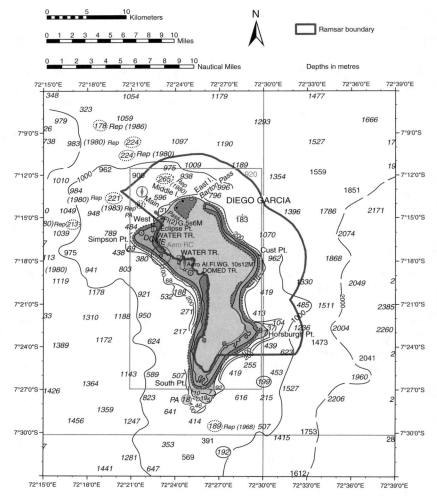

Map 3 Ramsar Convention Map—Diego Garcia Protected Site, 2001[82]

and the United Kingdom, which are among the world leaders in CO_2 emissions *per capita,* and hence primarily accountable for these changes,[87] may eventually have to convert the Empire's most important "strategic island" into an underwater archaeological site. In the long run, Diego Garcia will leave no footprint—not even a Cheshire Cat's grin on the face of the Indian Ocean.[88]

CHAPTER 6

Epilogue: The Lords' Day?

On October 22, 2008, the House of Lords' Appellate Committee—known as the "Law Lords"—by a 3:2 majority judgment allowed the UK government's appeal against the earlier British court decisions that had held the expulsion of the Chagossian islanders unlawful *(The Queen, on the application of Bancoult, vs. Secretary of State for Foreign and Commonwealth Affairs)*.[1] Lord Hoffmann, in his opinion for the majority (see appendix X), affirmed the lawfulness of the 2004 British Indian Ocean Territory (BIOT) Immigration Ordinance—denying the Chagossians a "right of abode" in their home islands—on the basis of the Colonial Laws Validity Act of 1865;[2] Lord Rodger of Earlsferry and Lord Carswell concurred, while Lord Bingham of Cornhill and Lord Mance dissented.[3]

So the Empire has struck back.[4] As a "conquered or ceded colony,"[5] the BIOT was held subject to the exclusive "prerogative" law-making powers of the British Crown (i.e., in practice, the cabinet acting through the Foreign and Commonwealth Office) without *any* parliamentary representation or participation.[6] While theoretically conceding the possibility of judicial review,[7] the Law Lords' majority opinion immediately carved out as "non-justiciable" the entire range of political decision-making relating to (1) military security and (2) budgetary matters.[8] On the former, they exempted from judicial scrutiny the government's reliance on what Lord Bingham's dissent described as "highly imaginative letters written by American officials to strengthen the Secretary of State's hand in this litigation" (raising the specter of terrorist risks to the security of Diego Garcia).[9] On the latter, they deferred to what the FCO described as an "independent study" commissioned by it,[10] raising the other specter of "prohibitive" future costs facing British taxpayers in the event of a BIOT resettlement[11]—which make the estimated legal costs of over £2.6 million (approximately $4 million) already incurred for this case by the UK government almost sound like a prudent investment.[12]

The majority then went on to reject as "extreme" the plaintiffs' (and the dissenting opinions') assertion of a fundamental human right of abode in one's own country, alleged to go back to the Magna Carta,[13] and concluded that the right so alleged cannot trump "primary" colonial legislation such as the 2004 BIOT Immigration Ordinance.[14] In the end, the Law Lords' majority opinion comes congenially close indeed to the related earlier U.S. judgment in *Olivier Bancoult et al. vs. Robert S. McNamara et al.*,[15] holding that the Chagossians's claims did not involve "constitutionally protected liberties" and declining judicial review of "political questions" reserved to the wisdom of the executive.[16]

The outcome does not make the whole Chagos a "legal black hole" *à la* Guantánamo;[17] rather, its legal system may now qualify as a "gray hole," governed by a kind of "British law light." As Lord Hoffmann expressly put it, the UK Human Rights Act of 1998 does *not* apply to BIOT because Britain has not formally extended her ratification of the 1950 European Human Rights Convention to the territory.[18] What he failed to mention was that, by the same token, the 1949 Geneva Conventions III and IV,[19] the 1966 UN Human Rights Covenants, the 1984 UN Convention Against Torture, the 1987 European Convention for the Prevention of Torture, and the 1998 Statute of the International Criminal Court also do *not* apply in BIOT—let alone other multilateral instruments such as the 1992 Biodiversity Convention or the 1998 Aarhus Convention on Access to Environmental Information, Justice and Decision-Making.[20] It is true, of course, as pointed out by Lord Rodger of Earlsferry for the majority, that the BIOT Courts Ordinance of 1983 (in force since February 1, 1984) provides that the law of the territory is "the law of England as from time to time in force,"

> provided that the said law of England shall apply in the Territory only so far as it is applicable and suitable to local circumstances, and shall be construed with such modifications, adaptations, qualifications and exceptions as local circumstances render necessary. (Section 3)[21]

The question is whether the law of England *minus* the Magna Carta still is the law of England. That question, however, may well prove irrelevant once the Chagossians's case moves to the European Court of Human Rights in Strasbourg.[22]

Diego Garcia's tragedy has been that both Britain and the United States consider the island a "fortress territory" where decolonization is "unthinkable."[23] One of the constitutional points repeatedly emphasized by the majority was the principle that, in a "ceded colony" such as BIOT, Her Majesty in Council is entitled to legislate in the interests of the United Kingdom, and "in the event of a conflict of interest, she is entitled, on the advice of Her United Kingdom ministers, to prefer the interests of the United Kingdom [over the interests of the Chagossians]."[24] That surely is a far cry from the "sacred trust of civilization" for non-self-governing territories and peoples in Article 73 of the UN Charter, which unequivocally declares "the principle that the interests of the inhabitants of these territories are paramount."[25] The sacred-trust concept, be it recalled, was initially coined in 1919 by President Woodrow Wilson.[26] By 1945 already, according to international law historians, "the doctrine of the 'sacred trust of civilization' had replaced formal European imperialism as the perspective from which international law conceived Europe's outside."[27] Or—has it now?

In the wake of *Bancoult 2*, a consortium of British and American conservation charities, established in 2008 as "Chagos Environment Network," has proposed to turn the entire archipelago (albeit minus Diego Garcia) into a permanent marine reserve,[28] following the example of the U.S. "marine national monuments" created by former president George W. Bush in the Pacific Ocean on January 6, 2009.[29] Laudable as the ecological intentions of such a proposal may be, it raises at least three major problems:

1. **Perpetuation of a colonial regime of governance for BIOT would violate the rights of the Chagos islanders.**
   There was no prior consultation with representatives of the indigenous Chagossians in exile who were expelled from the islands forty years ago.[30] Benevolent suggestions to the effect that some of the former inhabitants might be allowed to return to their home as "environmental wardens"—that is, as service-level employees of the new Anglo-American master regime for a park—are nothing short of cynical. If the governance institutions of the proposed reserve are to have any semblance of legitimacy, they would clearly have to include the Chagossians

themselves in decision-making functions.[31] Furthermore, in order to prevent their future discriminatory treatment and a return to the *de facto* slavery status of their forefathers under colonial rule,[32] the Foreign and Commonwealth Office would finally have to extend to BIOT all relevant British legislation and ratified treaties regarding human rights, as well as the Aarhus Convention.[33]

2. **Permanent designation of the Chagos islands as a British marine reserve would contravene the UK government's commitment to cede the islands to Mauritius "when they are no longer needed for defence purposes."**

 Mauritius has never given up its claim to sovereignty over "the Chagos Archipelago including Diego Garcia" (as stated in article 111 of its Constitution).[34] The European Union has formally recognized Mauritian jurisdiction over the 200-mile exclusive economic zone around the Chagos archipelago since 1989.[35] All successive British governments since 1983 have declared and reiterated their firm commitment to cede the Chagos islands to Mauritius "when they are no longer needed for defence purposes."[36] Following public statements by the chairman of the UK Chagos Conservation Trust claiming that the establishment of the Chagos archipelago as a marine reserve has now become "compatible with defence,"[37] it may be inferred that most of the islands are no longer needed for defence purposes—in which case they should rightfully revert to Mauritius to begin with.

 Nothing of course could stop Mauritius even now from reporting on the Chagos as part of its next national report to the 1992 Biodiversity Convention; or from declaring the Chagos a marine protected area of its own under Article 8 of the 1985 Protocol concerning Protected Areas and Wild Fauna and Flora in the Eastern African Region (whose bureau is chaired by Mauritius, and whose next conference of parties will be held in Port Louis in 2010),[38] or in the context of the Indian Ocean Commission's regional network of marine protected areas (headquartered in Mauritius).[39] A possible way out of the British-Mauritian diplomatic dilemma over the Chagos might indeed be the approach successfully followed under Article IV(1) of the 1959 Antarctic Treaty,[40] which by mutual consent "froze" all conflicting claims of territorial sovereignty in the area—though not, as the FCO and the State Department would have it, for the purpose of evading a multilateral treaty,[41] but rather in order to ensure effective joint application of the treaty.

3. **Unilateral enclosure of the Chagos 200-mile zone as a national nature protection area would not be borne out by the UN Convention on the Law of the Sea.**

The British Indian Ocean Territory, including its 3-mile territorial waters, covers an area of approximately 21,000 square miles (54,400 square kilometers).[42] The area now envisaged for inclusion in the proposed mega-marine park, however, is given as about 250,000 square miles (647,500 square kilometers,[43] equivalent to the size of Afghanistan); that is, the entire 200-mile BIOT "Environment (Protection and Preservation) Zone" unilaterally declared by the UK on 17 September 2003.[44] Besides being in open conflict with the competing/overlapping exclusive economic zones of Mauritius and the Maldives,[45] that zone has no formal legal basis in the 1982 UN Convention on the Law of the Sea (UNCLOS). During UNCLOS negotiations, a Canadian proposal to authorize coastal countries to establish a 100-mile environmental protection zone was opposed (not least by the U.S.) as an encroachment on the customary freedom of navigation,[46] and was rejected except as regards jurisdiction over ice-covered areas (not an immediate prospect in the Chagos).[47] Subsequent attempts at "green enclosure" of ocean areas—for example, the "ecological protection zones" unilaterally established by France, Croatia and Italy in the Mediterranean in 2003 and 2006—have remained controversial.[48] Significantly, when the U.S. established a 50-mile marine environment protection area surrounding its Northwestern Hawaiian Islands in 2006, it sought and obtained multilateral approval by the International Maritime Organization (IMO),[49] and applied for registration of the area under the World Heritage Convention.[50] If a similar status were now envisaged for the Chagos archipelago, the FCO would be expected to take similar steps on the basis of intergovernmental consultations, inevitably including the consent of Mauritius and the Maldives.

Instead of succumbing to the neo-Seldenian "territorial temptation" to nationalize yet another slice of ocean space,[51] the proponents of the new mega-marine park would be well advised to consider an alternative approach; viz., *public trusteeship*—a concept well established not only in the context of decolonization,[52] but also as an effective instrument of modern environmental law,[53] which may indeed be a viable solution for the sustainable management of marine resources in the exclusive economic zone.[54] The BIOT Administration would thus be held accountable as trustee—rather than "owner"—of the global natural heritage of the Chagos. The message is simple: The rights of nation states over ocean resources are never proprietary, but fiduciary.[55]

Appendices

U.S.-UK Official Documents on Diego Garcia

I. Agreement London, December 1966[1]
Availability for Defence Purposes of the British Indian Ocean Territory

I

Note No. 25
From David K.E. Bruce,
Ambassador of the United States of
America, London

December 30, 1966

To the Right Honorable George Brown, M.P.,
Secretary of State for Foreign Affairs
Foreign Office, London

Sir,

I have the honor to refer to recent discussion between representatives of the Government of the United States of America and the Government of the United Kingdom of Great Britain and Northern Ireland concerning the availability, for the defense purposes of both Governments as they may arise, of the islands of Diego Garcia and the remainder of the Chagos Archipelago, and the islands of Aldabra, Farquhar, and Desroches, constituting the British Indian Ocean Territory, hereinafter referred to as "the Territory." The United

States Government has now authorized me to propose an Agreement in the following terms:

1. The Territory shall remain under United Kingdom sovereignty.
2. Subject to the provisions set out below, the islands shall be available to meet the needs of both Governments for defense. In order to ensure that the respective United States and United Kingdom defense activities in the islands are correlated in an orderly fashion:
    (a) In the case of the initial United States requirement for use of a particular island, the appropriate governmental authorities shall consult with respect to the time required by the United Kingdom authorities for taking those administrative measures that may be necessary to enable such defense requirement to be met.
    (b) Before either Government proceeds to construct or install any facility in the Territory, both Governments shall first approve in principle the requirement for that facility, and the appropriate administrative authorities of the two Governments shall reach mutually satisfactory arrangements concerning specific area and technical requirements for respective defense purposes.
    (c) The procedure described in sub-paragraphs *(a)* and *(b)* shall not be applicable in emergency circumstances requiring temporary use of an island or part of an island not in use at that time for defense purposes provided that measures to ensure the welfare of the inhabitants are taken to the satisfaction of the Commissioner of the Territory. Each Government shall notify the other promptly of any emergency requirements and consultation prior to such use by the United States Government shall be undertaken as soon as possible.
3. The United Kingdom Government reserves the right to permit the use by third countries of British-financed defense facilities, but shall where appropriate consult with the United States Government before granting such permission. Use by a third country of United States or jointly-financed facilities shall be subject to agreement between the United Kingdom Government and the United States Government.
4. The required sites shall be made available to the United States authorities without charge.
5. Each Government shall normally bear the cost of site preparation, construction, maintenance, and operation for any facilities

developed to meet its own requirements. Within their capacities, such facilities shall be available for use by the forces of the other Government under service-level arrangements. However, there may be certain cases where joint financing should be considered, and these cases the two Governments shall consult together.

6. Commercial aircraft shall not be authorized to use military airfields in the Territory. However, the United Kingdom Government reserves the right to permit the use in exceptional circumstances of such airfields, following consultation with the authorities operating the airfields concerned, under such terms or conditions as may be defined by the two Governments.

7. For its defense purposes on the islands, the United States Government may freely select United States contractors and the sources of equipment, materials, supplies, or personnel except that—
   (a) the United States Government and United States contractors shall make use of workers from Mauritius and Seychelles to the maximum extent practicable, consistent with United States policies, requirements and schedules; and
   (b) the appropriate administrative authorities of the two Governments shall consult before contractors or workers from a third country are introduced.

8. The exemption from charges in the nature of customs duties and other taxes in respect of goods, supplies and equipment brought to the Territory in connection with the purposes of this Agreement by or on behalf of the United States Government, United States contractors, members of the United States Forces, contractor personnel or dependents, and the exemption from taxation of certain persons serving or employed in the Territory in connection with those purposes, shall be such exemption as is set out in Annex I to this Note.

9. The arrangements regarding the exercise of criminal jurisdiction and claims shall be those set out in Annex II to this Note.

10. For the purpose of this Agreement:
    (a) "Contractor personnel" means employees of a United States contractor who are not ordinarily resident in the Territory and who are there solely for the purpose of this Agreement;
    (b) "Dependents" means the spouse and children under 21 years of age of a person in relation to whom it is used; and, if they are dependent upon him for their support, the parents and children over 21 years of age of that person;

- **(c)** "Members of the United States Forces" means
  - **(i)** military members of the United States Forces on active duty;
  - **(ii)** civilian personnel accompanying the United States Forces and in their employ who are not ordinarily resident in the Territory and who are there solely for the purpose of this Agreement; and
  - **(iii)** dependents of the persons described in *(i)* and *(ii)* above;
- **(d)** "United States authorities" means the authority or authorities from time to time authorized or designated by the United States government for the purpose of exercising the powers in relation to which the expression is used;
- **(e)** "United States contractor" means any person, body or corporation ordinarily resident in the United States of America, that, by virtue of contract with the United States Government, is in the Territory for the purposes of this Agreement, and includes a sub-contractor;
- **(f)** "United States Forces" means the land, sea and air armed services of the United States, including the Coast Guard.

11. The United States Government and the United Kingdom government contemplate that the islands shall remain available to meet the possible defense needs of the two Governments for an indefinitely long period. Accordingly, after an initial period of 50 years this Agreement shall continue in force for a further period of twenty years unless, not more than two years before the end of the initial period, either Government shall have given notice of termination to the other, in which case this Agreement shall terminate two years from the date of such notice.

If the foregoing proposal is acceptable to the Government of the United Kingdom of Great Britain and Northern Ireland, I have the honor to propose that this Note and its Annexes, together with your reply to that effect, shall constitute an Agreement between the two Governments which shall enter into force on the date of your reply.

Accept, Sir, the renewed assurances of my highest consideration.

[signed: *David Bruce*]
American Ambassador

II

From the Minister of State for Foreign Affairs   30 December 1966
Foreign Office, London

To H.E. The Honourable David K.E. Bruce CBE,
Ambassador of the United States of America
London

Your Excellency,

I have the honour to acknowledge receipt of your Note No. 25 of the 30th of December, 1966, which reads as follows:

[see Note *I* above]

I have the honour to inform Your Excellency that the foregoing proposal is acceptable to the Government of the United Kingdom of Great Britain and Northern Ireland, who therefore agree that Your Excellency's Note, together with the Annexes thereto and this reply, shall constitute an Agreement between the two Governments which shall enter into force on this day's date.

I have the honour to be, with the highest consideration, Your Excellency's obedient Servant,

For the Secretary of State:
[signed: *Chalfont*]

– – –

Annex I: Customs Duties and Taxation

1. *Customs Duties and Other Taxes on Goods*
   (1) No import, excise, consumption or other tax, duty or impost shall be charged on:
      (a) material, equipment, supplies, or goods for use in the establishment, maintenance, or operation of the facilities which are consigned to or destined for the United States authorities or a United States contractor;
      (b) goods for the use or consumption aboard United States public vessels or aircraft;
      (c) goods consigned to the United States authorities or to a United States contractor for the use or for sale to military members of the United States Forces, or to other members

of the United States Forces, or to those contractor personnel and their dependents who are not engaged in any business or occupation in the Territory;
- (d) the personal belongings or household effects for the personal use of persons referred to in sub-paragraph *(c)* above, including motor vehicles, provided that these accompany the owner or are imported:
  - (i) within a period beginning sixty days before and ending 120 days after the owner's arrival; or
  - (ii) within a period of six months immediately following his arrival.
- (e) goods for consumption and goods (other than personal belongings and household effects) acquired after first arrival, including gifts, consigned to military members of the United States Forces, or to those other members of the United States Forces who are nationals of the United States and are not engaged in any business or occupation in the Territory, provided that such goods are:
  - (i) of United States origin if the Commissioner so requires, and
  - (ii) imported for the personal use of the recipient.
(2) No export tax shall be charged on the material, equipment, supplies or goods mentioned in paragraph *(1)* in the event of reshipment from the Territory.
(3) Article 1 of this Annex shall apply notwithstanding that the material, equipment, supplies or goods pass through other parts of the Territory en route to or from a site.
(4) The United States authorities shall do all in their power to prevent any abuse of customs privileges and shall take administrative measures, which shall be mutually agreed upon between the appropriate authorities of the United States and the Territory, to prevent the disposal, whether by resale or otherwise, of goods which are used or sold under paragraph *(1)(c)*, or imported under paragraph *(1)(d)* or *(1)(e)*, of Article 1 of this Annex, to persons not entitled to buy goods pursuant to paragraph *(1)(c)*, or not entitled to free importation under paragraph *(1)(d)* or *(1)(e)*. There shall be cooperation between the United States authorities and the Commissioner to this end, both in prevention and in investigation of cases of abuse.

2. *Motor Vehicle Taxes*

No tax or fee shall be payable in respect of registration or licensing for use for the purpose of this Agreement in the Territory or motor vehicles belonging to the United States Government or United States contractors.

3. *Taxation*
   (1) No members of the United States Forces, or those contractor personnel and their dependents who are nationals of the United States, serving or employed in the Territory in connection with the facilities shall be liable to pay income tax in the Territory except in respect of income derived from activities within the Territory other than such service or employment.
   (2) No such person shall be liable to pay in the Territory any poll tax or similar tax on his person, or any tax on ownership or use of property which is situated outside the Territory or situated within the Territory solely by reason of such person's presence there in connection with activities under this Agreement.
   (3) No United States contractor shall be liable to pay income tax in the Territory in respect of any income derived under a contract made in the United States in connection with the purposes of this Agreement, or any tax in the nature of license in respect of any service or work for the United States Government in connection with the purposes of this Agreement.

Annex II: Jurisdiction and Claims

1. *(a)* Subject to the provisions of sub-paragraphs *(b)* to *(l)* of this paragraph,
   - *(i)* the military authorities of the United States shall have the right to exercise within the Territory all criminal and disciplinary jurisdiction conferred on them by United States law over all persons subject to the military law of the United States; and
   - *(ii)* the authorities of the Territory shall have jurisdiction over the members of the United States Forces with respect to offenses committed within the Territory and punishable by the law in force there.

   (b) *(i)* The military authorities of the United States shall have the right to exercise exclusive jurisdiction over persons subject to the military law of the United States with respect to offenses, including offenses relating to security, punishable by the law of the United States but not by the law in force in the Territory.

  (ii) The authorities of the Territory shall have the right to exercise exclusive jurisdiction over members of the United States Forces with respect to offenses, including offenses relating to security, punishable by the law in force in the Territory but not by the law of the United States.
  (iii) For the purposes of sub-paragraphs *(b)* and *(c)*, an offense relating to security shall include:
   (aa) treason; and
   (bb) sabotage, espionage or violation of any law relating to official secrets or secrets relating to national defense.
(c) In cases where the right to exercise jurisdiction is concurrent, the following rules shall apply:
  (i) The military authorities of the United States shall have the primary right to exercise jurisdiction over a member of the United States Forces in relation to
   (aa) offenses solely against the property or security of the United States or offenses solely against the person or property of another member of the United States Forces; and
   (bb) offenses arising out of any act or omission done in the performance of official duty.
  (ii) In the case of any other offense, the authorities of the Territory shall have the primary right to exercise jurisdiction.
  (iii) If the authorities having the primary right decide not to exercise jurisdiction, they shall notify the other authorities as soon as practicable. The United States authorities shall give sympathetic consideration to a request from the authorities of the Territory for a waiver of their primary right in cases where the authorities of the Territory consider such waiver to be of particular importance. The authorities of the Territory will waive, upon request, their primary right to exercise jurisdiction under this paragraph, except where they in their discretion determine and notify the United States authorities that it is of particular importance that such jurisdiction be not waived.
(d) The foregoing provisions of this paragraph shall not imply any right for the military authorities of the United States to exercise jurisdiction over persons who belong to, or are ordinarily resident in, the Territory, or who are British subjects or

Commonwealth citizens or British protected persons, unless they are military members of the United States Forces.

(e) (i) To the extent authorized by law, the authorities of the Territory and the military authorities of the United States shall assist each other in the service of process and in the arrest of members of the United States Forces in the Territory and in handing them over to the authorities which are to exercise jurisdiction in accordance with provisions of this paragraph.

(ii) The authorities of the Territory shall notify promptly the military authorities of the United States of the arrest of any member of the United States Forces.

(iii) Unless otherwise agreed, the custody of an accused member of the United States Forces over whom the authorities of the Territory are to exercise jurisdiction shall, if he is in the hands of the United States authorities, remain with the United States authorities until he is charged. In cases where the United States authorities may have the responsibility for custody pending the completion of judicial proceedings, the United States authorities shall upon request make such a person immediately available to the authorities of the Territory for the purposes of investigation and trial and shall give full consideration to any special views of such authorities as to the way in which custody should be maintained.

(f) (i) To the extent authorized by law, the authorities of the Territory and of the United States shall assist each other in the carrying out of all necessary investigations into offenses, in providing for the attendance of witnesses and in the collection and production of evidence, including the seizure and, in proper cases, the handing over of objects connected with an offense. The handing over of such items may, however, be made subject to their return within the time specified by the authorities delivering them.

(ii) The authorities of the Territory and of the United States shall notify one another of the disposition of all cases in which there are concurrent rights to exercise jurisdiction.

(g) A death sentence shall not be carried out in the Territory by the military authorities of the United States.

(h) Where an accused has been tried in accordance with the provisions of this paragraph and has been acquitted or has been convicted and is serving, or has served, his sentence or has been pardoned, he may not be tried again for the same offense within the Territory. Nothing in this paragraph shall, however, prevent the military authorities of the United States from trying a military member of the United States for any violation of rules of discipline arising from an act or omission which constituted an offense for which he was tried by the authorities of the Territory.

(i) Whenever a member of the United States Forces is prosecuted by the authorities of the Territory, he shall be entitled
   (i) to a prompt and speedy trial;
   (ii) to be informed in advance of trial of the specific charge or charges made against him;
   (iii) to be confronted with the witnesses against him;
   (iv) to have compulsory process for obtaining witnesses in his favor if they are within the jurisdiction of the Territory;
   (v) to have legal representation of his own choice for his defense or to have free or assisted legal representation under conditions prevailing for the time being in the Territory;
   (vi) if he considers it necessary, to have the services of a competent interpreter; and
   (vii) to communicate with a representative of the United States and, when the rules of the court permit, to have such representative present at his trial which shall be public except when the court decrees otherwise in accordance with the law in force in the Territory.

(j) Where a member of the United States Forces is tried by the military authorities of the United States for an offense committed outside the areas used by the United States or involving a person, or the property of a person, other than a member of the United States Forces, the aggrieved party and representatives of the Territory and of the aggrieved party may attend the trial proceedings except where this would be inconsistent with the rules of the court.

(k) A certificate of the appropriate United States commanding officer that an offense arose out of an act or omission done in the performance of official duty shall be conclusive, but the commanding officer shall give consideration to any representation made by the authorities of the Territory.

(l) Regularly constituted military units or formations of the United States Forces shall have the right to police the areas used by the United States. The military police of the United States Forces may take all appropriate measures to ensure the maintenance of order and security within these areas.

2. (a) The Government of the United States of America and the Government of the United Kingdom respectively waive all claims against the other of them:

    (i) for damage to any property owned by it and used by its land, sea or air armed services if such damage

        (aa) was caused by a member of the armed services or by an employee of a Department with responsibilities for the armed services of either Government in the execution of his duties, or

        (bb) arose from the use of any vehicle, vessel or aircraft owned by either Government and used by its armed services provided either that the vehicle, vessel or aircraft causing the damage was being used in connection with official duties, or the damage was caused to property being so used;

    (ii) for injury or death suffered by any member of its armed services while such member was engaged in the performance of his official duties.

    (iii) For the purpose of this paragraph, "owned" in the case of a vessel includes a vessel on bare boat charter, a vessel requisitioned on bare boat terms and a vessel seized in prize (except to the extent that the risk of loss or liability is borne by some person other than either Government).

(b) (i) The United States Government shall, in consultation with the Government of the Territory, take all reasonable precautions against possible danger and damage resulting from operations under this Agreement.

    (ii) The United States Government agrees to pay just and reasonable compensation, which shall be determined in accordance with the measure of damage prescribed by the law of the Territory, in settlement of civil claims (other than contractual claims) arising out of acts or omissions of members of the United States Forces done in the performance of official duty or out of any other act or omission or occurrence for which the United States Forces are legally responsible.

(iii) Any such claim presented to the United States Government shall be processed and settled in accordance with the applicable provisions of United States law.

II. Agreed Confidential Minutes, London, December 1966[2]

Confidential

Agreed Minute

In the course of discussions leading up to the Exchange of Notes of 30 December 1966, constituting an Agreement between the Governments of the United States and the United Kingdom concerning the use of the islands in the British Indian Ocean Territory for defence purposes the following agreement and understandings were reached:

I With reference to paragraph **(2)(a)** of the Agreement, the administrative measures referred to are those necessary for modifying or terminating any economic activity then being pursued in the islands, resettling any inhabitants, and otherwise facilitating the availability of the islands for defence purposes.

Where any United States requirement is for land owned by the United Kingdom Government but in the possession of a lessee of that Government and it will be necessary for notice of termination of the lease to be given by or on behalf of that Government to the lessee, there will be adequate notice of the United States requirement for the purpose of enabling the United Kingdom Government to give the lessee six months' notice of the termination of the lease or such less period of notice as may be specified in the lease. This paragraph shall not, however, apply in the circumstances envisaged in paragraph **(2)(c)** of the Agreement.

II With reference to paragraph **(2)(b)** of the Agreement, the approval in principle by both Governments before either constructs or installs any facility is required only for construction or installation of major new developments. Such developments would be of the order of an air staging base, a fleet support installation, or a space tracking station. The mutually satisfactory arrangements between appropriate administrative authorities would be sufficient for improvement or reasonable expansion of approved facilities already constructed or installed.

III With reference to paragraph **(2)(c)** of the Agreement, the types of measures considered appropriate by the British authorities during periods of emergency use by the United States will be indicated to the United States authorities and will be reflected by the latter in any planning for emergency use. In the event of such emergency use of an inhabited island, the implementation of measures taken by the United States authorities to ensure the welfare of any inhabitants may be monitored by British personnel.

When temporary emergency use is required, the administrative authorities of the two Governments will agree upon the arrangements (if any) regarding such temporary use which may in the circumstances be appropriate.

IV With reference to paragraph **(6)** of the Agreement, the Governments of the United States and the United Kingdom agree to define the terms and conditions for use in exceptional circumstances by commercial aircraft of military airfields in the Territory, as follows:—

    (i) Such use shall be limited to technical stops by British and United States commercial aircraft only;

    (ii) the United States Government has indicated its agreement to such use following consultation on an expedited basis at the time, provided for in paragraph **(6)** of the Agreement, for the purpose of making practical arrangements;

    (iii) if, however, a third government should in the view of the United Kingdom make an effective challenge, in pursuance of international instruments relating to civil aviation, to the United Kingdom's action as sovereign power in denying the use of an airfield, then it is agreed that civil use by British and United States commercial aircraft shall be suspended for such time as in the view of the United Kingdom Government the effective challenge is maintained;

    (iv) the above provisions would not preclude the use of military airfields by civil aircraft operated by or on behalf of either Government for governmental purposes, which is covered by the service-level arrangements provided for in paragraph **(5)** of the Agreement.

V Paragraph **(2(b)(iii)** of **Annex II** to the Agreement does not debar any person who has a civil claim against the United States Government or any person for whose acts or omission that Government is responsible from bringing a civil claim in a British court under British law in any circumstances in which it would otherwise be open for him to do so.

VI In the light of circumstances prevailing in the Territory at the commencement of the Agreement and the use to which it is contemplated the islands will be put, no formal provision has been included to cover the status of the members of the United States Forces and other personnel (except with regard to jurisdiction, customs duties, and taxes) and certain defence activities of the United States pursuant to the Agreement. The lack of formal provisions in these respects will not operate to restrict such defence activities of the United States authorities. If at any time during the continuance of the Agreement it appears to the United Kingdom Government or the United States Government necessary, having regard to any change in the use of [*sic*; or?] any development in the circumstances of the islands, to make formal provision for those matters, an Agreement will be concluded containing such of the provisions of the Seychelles Tracking Facility Agreement[3] as appear necessary to the two Governments, with any necessary modifications, and such other provisions as appear necessary to the two Governments.

London, 30 December 1966   [signed: *David Bruce*]   [signed: *Chalfont*]

III. Supplement London, October 1972[4]

Limited United States Naval Communications Facility on Diego Garcia, British Indian Ocean Territory

I

Note No. HKT 10/1   24 October 1972

From the Secretary of State for Foreign and Commonwealth Affairs
Foreign and Commonwealth Office, London

To the Chargé d'affaires ad interim
Embassy of the United States of America, London

Sir,

I have the honour to refer to the Agreement constituted by the exchange of notes dated 30 December 1966, between the Government of the United Kingdom of Great Britain and Northern Ireland and the Government of the United States of America concerning the availability of the British Indian Ocean Territory for defence purposes. Pursuant to paragraph 2(b) of that Agreement, I now convey the approval in principle of the Government of the United Kingdom to the construction of a limited naval communications facility on Diego Garcia and propose an agreement in the following terms:

1. Scope of the facility

(a) Subject to the following provisions of this Agreement, the Government of the United States shall have the right to construct, maintain and operate a limited naval communications facility on Diego Garcia. The facility shall consist of transmitting and receiving services, an anchorage, airfield, associated logistic support and supply and personnel accommodation. For this purpose immovable structures, installations and buildings may be constructed within the specific area shown in the plan annexed to this note. The specific area may be altered from time to time as may be agreed by the appropriate administrative authorities of the two Governments.

(b) During the term of this Agreement the Government of the United States may conduct on Diego Garcia such functions as are necessary for the construction, maintenance, operation and security of the facility. For this purpose the Government of the

United States shall have freedom of access to that part of Diego Garcia outside the specific area referred to in sub-paragraph *(a)*, but may erect or construct immovable structures, installations and buildings outside the specific area only with the prior agreement of the appropriate administrative authorities of the Government of the United Kingdom.

(c) Delimitation of the specific area shall, subject to the provisions of the BIOT Agreement, in no way restrict the Government of the United Kingdom from constructing and operating their own defence facility within that area, provided that no technical interference to existing operations will result from such construction and operation.

2. Purpose

The facility shall provide a link in United States defence communications and shall furnish improved communications support in the Indian Ocean for ships and state aircraft owned or operated by or on behalf of either Government.

3. Access to Diego Garcia

(a) Access to Diego Garcia shall in general be restricted to members of the Forces of the United Kingdom and of the United States, the Commissioner and public officers in the service of the British Indian Ocean Territory, representatives of the Governments of the United Kingdom and of the United States and, subject to normal immigration requirements, contractor personnel. The Government of the United Kingdom reserves the right, after consultation with the appropriate United States administrative authorities, to grant access to members of scientific parties wishing to carry out research on Diego Garcia and its environs, provided that such research does not unreasonably interfere with the activities of the facility. The Commanding Officer shall afford appropriate assistance to members of these parties to the extent feasible and on a reimbursable basis. Access shall not be granted to any other person without prior consultation between the appropriate administrative authorities of the two Governments.

(b) Ships and state aircraft owned or operated by or on behalf of either Government may freely use the anchorage and airfield.

(c) Pursuant to the provisions of the second sentence of paragraph (3) of the BIOT Agreement, ships and state aircraft owned or

operated by or on behalf of a third government, and the personnel of such ships and aircraft, may use only such of the services provided by the facility, and on such terms, as may be agreed in any particular case by the two Governments.

4. Protection and security

Responsibility for protection and security of the facility shall be vested in the Commanding Officer, who shall maintain a close liaison with the Commissioner. The two Governments shall consult if there is any threat to the facility.

5. Shipping, navigation and aviation facilities

The Government of the United States shall have the right to install, operate and maintain on Diego Garcia such navigational and communications aids as may be necessary for the safe transit of ships and aircraft into and out of Diego Garcia.

6. Radio frequencies and telecommunications

(a) Subject to the prior concurrence of the Government of the United Kingdom, the Government of the United States may use any radio frequencies, powers and band widths for radio services (including radar) on Diego Garcia which are necessary for the operation of the facility. All radio communications shall comply at all times with the provisions of the International Telecommunications Convention.[5]

(b) The Government of the United States may establish such land lines on Diego Garcia as may be necessary for the facility.

7. Conservation

As far as possible the activities of the facility and its personnel shall not interfere with the flora and fauna of Diego Garcia. When their use is no longer required for the purposes of the facility, the two Governments shall consult about the condition of the three islets at the mouth of the lagoon with a view to restoring them to their original condition. However, neither Government shall be under any obligation to provide funds for such restoration.

8. Anchorage dues and aviation charges

Collection of dues and charges for use of the anchorage and airfield at Diego Garcia which may be levied by the Commissioner shall be his responsibility. State aircraft and ships owned or operated by or on behalf of the Government of the United States shall be permitted to use the anchorage and airfield without the payment of any dues or charges.

9. Meteorology

The Government of the United States shall operate a meteorological facility on Diego Garcia and supply such available meteorological information as may be required by the Government of the United Kingdom and the Government of Mauritius to meet their national and international obligations.

10. Royal Navy element

The Royal Navy element on Diego Garcia shall be under the command of a Royal Navy officer who shall be known as the Officer-in-Charge of the Royal Navy element. He shall be the Representative on Diego Garcia of the Commissioner.

11. Finance

The Government of the United States shall wholly bear the cost of constructing, operating and maintaining the facility. The Government of the United Kingdom shall be responsible for the pay, allowances and any other monetary gratuities of Royal Navy personnel, for the cost of their messing, and for supplies or services which are peculiar to or provided for the exclusive use of the Royal Navy or its personnel and which would not normally be provided by the Government of the United States for the use of its own personnel.

12. Fisheries, oil and mineral resources

The Government of the United Kingdom will not permit commercial fishing in the lagoon or oil or mineral exploration or exploitation on Diego Garcia for the duration of this Agreement. Furthermore, the Government of the United Kingdom will not permit commercial fishing or oil or mineral exploration or exploitation in or under those areas of the waters, continental shelf and sea-bed around Diego Garcia over which the United Kingdom has sovereignty or exercises sovereign rights, unless it is agreed that such activities would not harm or be inimical to the defence use of the island.

13. Health, quarantine and sanitation

The Commanding Officer and the Commissioner shall collaborate in the enforcement on Diego Garcia of necessary health, quarantine and sanitation provisions.

14. News broadcast station

The Government of the United States may establish and operate a closed circuit TV and a low power radio broadcast station to broadcast news, entertainment and educational programmes for personnel on Diego Garcia.

15. Property

(a) Title to any movable property brought into Diego Garcia by or on behalf of the Government of the United States, or by a United States contractor, shall remain in the Government of the United States or the contractor, as the case may be. Such property of the Government of the United States, including official papers, shall be exempt from inspection, search and seizure. Such property of either the Government of the United States or of a United States contractor may be freely removed from Diego Garcia, but shall not be disposed of within the British Indian Ocean Territory or Seychelles unless an offer, consistent with the laws of the United States then in effect, has been made to sell the property to the Commissioner and he has not accepted such offer within a period of 120 days after it was made or such longer period as may be reasonable in the circumstances. Any such property not removed or disposed of within a reasonable time after termination of this Agreement shall become the property of the Commissioner.

(b) The Government of the United States shall not be responsible for restoring land or other immovable property to its original condition, nor for making any payment in lieu of restoration.

16. Availability of funds

To the extent that the carrying out of any activity or the implementation of any part of this Agreement depends upon funds to be appropriated by the Congress of the United States, it shall be subject to the availability of such funds.

17. Restriction of rights

The Government of the United States shall not exercise any of the above rights or powers, or permit the exercise thereof, except for the purposes herein specified.

18. Supplementary arrangements

Supplementary arrangements between the appropriate administrative authorities of the two Governments may be made from time to time as required for the carrying out of the purposes of this Agreement.

19. Definitions and interpretation
   (a) For the purposes of this Agreement
      "BIOT Agreement" means the Agreement referred to in the first paragraph of this note;
      "Commanding Officer" means the United States Navy Officer in command of the facility;
      "Commissioner" means the officer administering the Government of the British Indian Ocean Territory;
      "Diego Garcia" means the atoll of Diego Garcia, the lagoon and the three islets at the mouth of the lagoon.
   (b) The provisions of this Agreement shall supplement the BIOT Agreement and shall be construed in accordance with that Agreement. In the event of any conflict between the provisions of the BIOT Agreement and this Agreement the provisions of the BIOT Agreement shall prevail.

20. Duration and termination

This Agreement shall continue in force for as long as the BIOT Agreement continues in force or until such time as no part of Diego Garcia is any longer required for the purposes of the facility, whichever occurs first.

If the Government of the United States of America also approves in principle the construction of the facility subject to the above terms, I have the honour to propose that this note and the plan annexed to it, together with your reply to that effect, shall constitute an agreement between the two Governments which shall enter into force on the date of your reply and shall be known as the Diego Garcia Agreement 1972.

I have the honour to be with high consideration, Sir,

For the Secretary of State:

[ANNEX: Diego Garcia Map/Plan 1972]   [signed: *Anthony Kershaw*]

II

No. 22 October 24, 1972

From the Chargé d'affaires ad interim
Embassy of the United States of America, London

To the Secretary of State for Foreign and Commonwealth Affairs
Foreign and Commonwealth Office, London

Sir,

I have the honor to acknowledge receipt of your note No. HKT 10/1 of October 24, 1972, which reads as follows:

[see Note I above]

I have the honor to inform you that the Government of the United States of America approves in principle the construction of the facility subject to the terms set out in your note, and therefore agree that your note, and the plan annexed to it, together with this reply, shall constitute an agreement between the two Governments which shall enter into force on today's date and shall be known as the Diego Garcia Agreement 1972.

Accept, Sir, the renewed assurances of my highest consideration.

[signed: *Earl D. Sohm*]

IV. Supplement London, February 1976[6]

United States Naval Support Facility on Diego Garcia, British Indian Ocean Territory

I

Note No. DPP 063/530/2   25 February 1976

From the Minister of State for Foreign and Commonwealth Affairs
Foreign and Commonwealth Office, London

To the Chargé d'affaires ad interim, Embassy of the United States of America, London

Sir,

I have the honour to refer to the Agreement constituted by the Exchange of Notes dated 30 December 1966 between the Government of the United Kingdom of Great Britain and Northern Ireland and the Government of the United States of America concerning the availability of the British Indian Ocean Territory for defence purposes and to the Agreement constituted by the Exchange of Notes dated 24 October 1972 between the two Governments concerning a limited United States naval communications facility on Diego Garcia, British Indian Ocean Territory. Pursuant to paragraph 2(b) of the former Agreement, I now convey the approval in principle of the Government of the United Kingdom to the development of the present limited naval communications facility on Diego Garcia into a support facility of the United States Navy and propose an Agreement in the following terms:

1. *Scope of the facility*
   (a) Subject to the following provisions of this Agreement, the Government of the United States shall have the right to develop the present limited naval communications facility on Diego Garcia as a support facility of the United States Navy and to maintain and operate it. The facility shall consist of an anchorage, airfield, support and supply elements and ancillary services, personnel accommodation, and transmitting and receiving services. Immovable structures, installations and buildings for the facility may, after consultation with the appropriate administrative authorities of the United Kingdom, be constructed

within the specific area shown in the plan attached to this Note. The specific area may be altered from time to time as may be agreed by the appropriate administrative authorities of the two Governments.

(b) During the term of the Agreement the Government of the United States may conduct on Diego Garcia such functions as are necessary for the development, use, maintenance, operation and security of the facility. In the exercise of these functions the Government of the United States, members of the United States Forces and contractor personnel shall have freedom of access to that part of Diego Garcia outside the specific area referred to in sub-paragraph *(a)*, but the Government of the United States may erect or construct immovable structures, installations and buildings outside the specific area only with the prior agreement of the appropriate administrative authorities of the Government of the United Kingdom.

(c) Delimitation of the specific area shall, subject to the provisions of the BIOT Agreement and after consultation with the appropriate United States authorities with a view to avoiding interference with the existing use of the facility, in no way restrict the Government of the United Kingdom from constructing and operating at their own expense their own defence facilities within that area, or from using that part of Diego Garcia outside the specific area.

2. Purpose

The facility shall provide an improved link in United States defence communications, and furnish support for ships and aircraft owned or operated by or on behalf of either Government.

3. Consultation

Both Governments shall consult periodically on joint objectives, policies and activities in the area. As regards the use of the facility in normal circumstances, the Commanding Officer and the Officer in Charge of the United Kingdom Service element shall inform each other of intended movements of ships and aircraft. In other circumstances the use of the facility shall be a matter for the joint decision of the two Governments.

4. Access to Diego Garcia

(a) Access to Diego Garcia shall in general be restricted to members of the Forces of the United Kingdom and of the United

States, the Commissioner and public officers in the service of the British Indian Ocean Territory, representatives of the Governments of the United Kingdom and of the United States and, subject to normal immigration requirements, contractor personnel. The Government of the United Kingdom reserves the right, after consultation with the appropriate United States administrative authorities, to grant access to members of scientific parties wishing to carry out research on Diego Garcia and its environs, provided that such research does not unreasonably interfere with the activities of the facility. The Commanding Officer shall afford appropriate assistance to members of these parties to the extent feasible and on a reimbursable basis. Access shall not be granted to any other person without prior consultation between the appropriate administrative authorities of the two Governments.

(b) Ships and aircraft owned or operated by or on behalf of either Government may freely use the anchorage and airfield.

(c) Pursuant to the provisions of the second sentence of paragraph (3) of the BIOT Agreement, ships and aircraft owned or operated by or on behalf of a third government, and the personnel of such ships and aircraft, may use only such of the services provided by the facility, and on such terms, as may be agreed in any particular case by the two Governments.

5. *Protection and security*

Responsibility for protection and security of the facility shall be vested in the Commanding Officer, who shall maintain a close liaison with the Commissioner. The two Governments shall consult if there is any threat to the facility.

6. *Shipping, navigation and aviation facilities*

The Government of the United States shall have the right to install, operate and maintain on Diego Garcia such navigational and communications aids as may be necessary for the safe transit of ships and aircraft into and out of Diego Garcia.

7. *Radio frequencies and telecommunications*

(a) Subject to the prior concurrence of the Government of the United Kingdom, the Government of the United States may use any radio frequencies, powers and band widths for radio services (including radar) on Diego Garcia which are necessary for the operation of the facility. All radio communications shall comply at all times with the provisions of the International Telecommunications Convention.

(b) The Government of the United States may establish such land lines on Diego Garcia as may be necessary for the facility.

8. Conservation

As far as possible the activities of the facility and its personnel shall not interfere with the flora and fauna of Diego Garcia. When their use is no longer required for the purposes of the facility, the two Governments shall consult about the condition of the three islets at the mouth of the lagoon with a view to restoring them to their original condition. However, neither Government shall be under any obligation to provide funds for such restoration.

9. Anchorage dues and aviation charges

Collection of dues and charges for use of the anchorage and airfield at Diego Garcia which may be levied by the Commissioner shall be his responsibility. Aircraft and ships owned or operated by or on behalf of the Government of the United States shall be permitted to use the anchorage and airfield without the payment of any dues or charges.

10. Meteorology

The Government of the United States shall operate a meteorological facility on Diego Garcia and supply such available meteorological information as may be required by the Government of the United Kingdom and the Government of Mauritius to meet their national and international obligations.

11. United Kingdom Service element

The United Kingdom Service element on Diego Garcia shall be under the Command of a Royal Navy Officer who shall be known as the Officer-in-Charge of the United Kingdom Service element.

12. Finance

(a) The Government of the United States shall bear the cost of developing, operating and maintaining the facility. However, in relation to United Kingdom personnel attached to the facility, the Government of the United Kingdom shall be responsible for their pay, allowances and any other monetary gratuities, for the cost of their messing, and for supplies or services which are peculiar to or provided for the exclusive use of the United

Kingdom Services or their personnel and which would not normally be provided by the Government of the United States for the use of its own personnel.

(b) Except in relation to the United Kingdom Service personnel attached to the facility, logistic support furnished at Diego Garcia by either Government, upon request, to the other Government, shall be on a reimbursable basis in accordance with the laws, regulations and instructions of the Government furnishing the support.

13. Fisheries, oil and mineral resources

The Government of the United Kingdom will not permit commercial fishing in the lagoon or oil or mineral exploration or exploitation on Diego Garcia for the duration of this Agreement. Furthermore, the Government of the United Kingdom will not permit commercial fishing or oil or mineral exploration or exploitation in or under those areas of the waters, continental shelf and sea-bed around Diego Garcia over which the United Kingdom has sovereignty or exercises sovereign rights, unless it is agreed that such activities would not harm or be inimical to the defence use of the island.

14. Health, quarantine and sanitation

The Commanding Officer and the Commissioner shall collaborate in the enforcement on Diego Garcia of necessary health, quarantine and sanitation provisions.

15. News broadcast station

The Government of the United States may establish and operate a closed circuit TV and a low power radio broadcast station to broadcast news, entertainment and educational programmes for personnel on Diego Garcia.

16. Property

(a) Title to any removable property brought into Diego Garcia by or on behalf of the Government of the United States, or by a United States contractor, shall remain in the Government of the United States or the contractor, as the case may be. Such property of the Government of the United States, including official papers, shall be exempt from inspection, search and seizure. Such property of either the Government of the United States or of a United States contractor may be freely removed from Diego Garcia, but shall not be disposed of within the British Indian Ocean Territory or Seychelles unless an offer, consistent with the laws of the United States then in effect, has been made to sell the property to the Commissioner and he has not accepted such offer within a period of 120 days after it was made or such longer period as may be reasonable in the circumstances. Any such property not removed or disposed of within a reasonable time after termination of this Agreement shall become the property of the Commissioner.

(b) The Government of the United States shall not be responsible for restoring land or other immovable property to its original condition, not for making any payment in lieu of restoration.

17. Availability of funds

To the extent that the carrying out of any activity or the implementation of any part of this Agreement depends upon funds to be appropriated by the Congress of the United States, it shall be subject to the availability of such funds.

18. Representative of the Commissioner

The Commissioner shall designate a person as his Representative on Diego Garcia.

19. Supplementary arrangements

Supplementary arrangements between the appropriate administrative authorities of the two Governments may be made from time to time as required for the carrying out of the purposes of this Agreement.

20. Definitions and interpretation

(a) For the purposes of this Agreement

"BIOT Agreement" means the Exchange of Notes dated 30 December 1966, between the Government of the United Kingdom of Great Britain and Northern Ireland and the Government of the United States of America concerning the availability of the British Indian Ocean Territory for defence purposes;

"Commanding Officer" means the United States Navy Officer in command of the facility; "Commissioner" means the officer administering the Government of the British Indian Ocean Territory;

"Diego Garcia" means the atoll of Diego Garcia, the lagoon and the three islets at the mouth of the lagoon.

(b) Questions of interpretation arising from the application of this Agreement shall be the subject of consultation between the two Governments.

(c) The provisions of this Agreement shall supplement the BIOT Agreement and shall be construed in accordance with that Agreement. In the event of any conflict between the provisions of the BIOT Agreement and this Agreement the provisions of the BIOT Agreement shall prevail.

21. The Diego Garcia Agreement 1972

This Agreement shall replace the Agreement constituted by the Exchange of Notes dated 24 October 1972 between the Government of the United Kingdom of Great Britain and Northern Ireland and the Government of the United States of America concerning a limited United States naval communication facility on Diego Garcia, British Indian Ocean Territory.

22. Duration and termination

This Agreement shall continue in force for as long as the BIOT Agreement continues in force or until such time as no part of Diego Garcia is any longer required for the purposes of the facility, whichever occurs first.

If the Government of the United States of America also approves in principle the development of the facility subject to the above terms, I have the honour to propose that this Note and the plan annexed to it, together with your reply to that effect, shall constitute an Agreement between the two Governments which shall enter into force on the date of your reply and shall be known as the Diego Garcia Agreement 1976.

I have the honour to be with high consideration, Sir,

Your obedient Servant,

[signed: *Roy Hattersley*]

[ANNEX: Diego Garcia Map/Plan 1976]

II

From the Chargé d'affaires ad interim
Embassy of the United States of America
London

25 February 1976

To the Minister of State for Foreign and Commonwealth Affairs
Foreign and Commonwealth Office
London

Sir,

I have the honor to acknowledge receipt of your Note No. DPP 063/530/2 of 25 February 1976, which reads as follows:

[see Note I above]

I have the honor to inform you that the Government of the United States of America approves in principle the development of the facility subject to the terms set out in your Note, and therefore agree that your Note, and the plan annexed to it, together with this reply, shall constitute an Agreement between the two Governments which shall enter into force on today's date and shall be known as the Diego Garcia Agreement 1976.

Accept, Sir, the renewed assurances of my highest consideration.

[signed: *Ronald I. Spiers*]

Supplementary Arrangements 1976
For Diego Garcia Facility[7]

Preamble

Pursuant to paragraph 19 of the Diego Garcia Agreement 1976 between the Government of the United Kingdom of Great Britain and Northern Ireland and the Government of the United States of America concerning the United States Navy support facility on Diego Garcia, the Ministry of Defence (Navy) and the United States Navy (USN) have made the following supplementary arrangements:

Paragraph 1: Personnel

The USN will establish a manning level for the facility. Representatives of both administrative authorities will jointly decide which positions shall be filled by UK Service personnel. All personnel assigned to Diego Garcia will serve an unaccompanied tour of duty.

Paragraph 2: Military Command

The Officer-in-Charge of the UK Service element will, in matters relating to the operation of the facility, report to the Commanding Officer. The Commanding Officer and the Officer-in-Charge of the UK Service element will establish the manner in which orders and instructions will be complied with, which manner will be consistent with the concept of mutual respect for relative ranks. However, nothing in this paragraph is intended to require obedience to any command inconsistent with the obligation of their respective service laws nor to establish disciplinary power in either officer over members of the Armed Services of the other country.

Paragraph 3: Logistic Support

Subject to Paragraph 4 below, military personnel of both Governments will be entitled to use: upon the same terms and conditions, such recreational, accommodation and messing facilities as are available or as are established for military personnel by either Government. UK Service personnel serving with this facility will be entitled to send and receive mail through the United States Fleet postal system. The USN will, upon request, transport UK Service personnel to and from the facility from such places as may be agreed from time to time by the USN and the Ministry of Defence of the United Kingdom (MOD). For the purpose of such transport UK Service personnel may be accompanied by personal baggage which does not exceed a gross weight of 120 pounds per man. The USN will give sympathetic consideration to requests for transportation of official UK Service visitors. The USN will, upon request, provide such supplies and services on an equivalent basis with USN personnel as may be required by UK Service personnel serving with the facility on Diego Garcia. When these supplies and services are peculiar to the UK Services the MOD will make them available to the USN at a place or places agreed to by the MOD and the USN at the time.

Paragraph 4: Finance

The financial arrangements have been laid down in paragraph 12 of the Diego Garcia Agreement 1976, which reads as follows:

(a) The Government of the United States shall bear the cost of developing, operating and maintaining the facility. However, in

relation to United Kingdom Service personnel attached to the facility, the Government of the United Kingdom shall be responsible for their pay, allowances and any other monetary gratuities, for the cost of their messing, and for supplies or services which are peculiar to or provided for the exclusive use of the United Kingdom Services or their personnel and which would not normally be provided by the Government of the United States for the use of its own personnel.

(b) Except in relation to UK Service Personnel attached to the facility logistic support furnished at Diego Garcia by either Government, upon request, to the other Government, shall be on a reimbursable basis in accordance with the laws, regulations and instructions of the Government furnishing the support.

Paragraph 5: Radio Frequencies and Telecommunications

The following procedures for obtaining the prior concurrence of the Government of the United Kingdom to the use of any radio frequencies, powers and band widths for radio services (including radar) on Diego Garcia which are necessary for the operation of the facility, and for international notification, will be followed:

a. Prior to the assignment, or modification of an assignment, of any radio frequency on Diego Garcia, concurrence for the same will be obtained from the United Kingdom through the established military co-ordination channel. This channel is between the Joint Frequency Panel (J/FP), USMCEB and the Defence Signal Staff, Signals 2 (DSS 2) Ministry of Defence, United Kingdom.

b. Upon obtaining such concurrence the United States will transmit to the International Frequency Registration Board (IFRB) notification of the assignment in accordance with existing US/UK frequency co-ordination procedures.

Paragraph 6: Aids to Navigation and Approach Control

The United States may use and maintain existing electronic navigation and landing aids, such as airport surveillance radar, ground controlled approach (GCA), Tacan and instrument landing systems (ILS). If in the future it should be necessary to make significant changes to the present electronic navigation and landing aids or to expand

Appendices

them significantly, this may be done subject to agreement between the MOD and the USN.

Paragraph 7: Scientific Research

If the Government of the United Kingdom wishes to grant access to Diego Garcia to members of scientific parties wanting to carry out research on Diego Garcia and its environs written notice will be given to the United States Department of State or the US Embassy in London at least four weeks prior to the intended visit. This notice will contain the following information:

a. identification of visiting party, including nationality and names of all members of the party;
b. scientific purpose;
c. date of arrival and expected duration;
d. areas to be utilised;
e. activities to be conducted;
f. equipment to be utilised;
g. services requested from the facility.

Such notice and the response thereto will constitute the consultation referred to in sub-paragraph 3(a) of the Diego Garcia Agreement 1976. Scientific parties will, where necessary, be responsible for reimbursing the Government of the United States for any goods and services supplied to them by the USN.

Paragraph 8: Local Administration

The following matters have been authorised by the Commissioner BIOT:

a. *Drivers' Licences.* United States or United Kingdom motor vehicle drivers' licences will be accepted as valid for the operation of all motor vehicles on Diego Garcia.
b. *Medical Services.* US medical personnel may perform medical services in Diego Garcia of the same type which such persons are authorised to perform at United States military medical facilities without prior examination or revalidation of their professional certificate by the United Kingdom authorities, and such facilities will be made available to United Kingdom Service personnel.

For the purposes of this paragraph, the term "US medical personnel" means the physicians, surgeons, specialists, dentists, nurses and other United States personnel in Diego Garcia who perform medical services, and other doctors of United States nationality or ordinarily resident in the United States employed or contracted in exceptional cases by the United States Forces.

c. *Recreational Fishing.* United States personnel and United Kingdom personnel are permitted reasonable recreational fishing on Diego Garcia and its environs without obtaining any licence or paying any fees. Such recreational fishing includes fishing from boats as well as from the shore.

Paragraph 9: Alteration

These Supplementary Arrangements may be altered at any time by the mutual consent of the parties hereto.

Paragraph 10: Interpretation

Unless the context otherwise requires, terms and expressions used herein will have the meanings assigned to them in the Diego Garcia Agreement 1976. In the event of any conflict between the provisions of these Supplementary Arrangements and of the Diego Garcia Agreement 1976 the latter will prevail.

For the Royal Navy:
[signed: *R.D. Lygo*]
Vice Admiral

For the United States Navy:
[signed: *D.H. Bagley*] Admiral

Signed in duplicate at London, the twenty-fifth day of February 1976.

Related Notes

I

No. 5   25 February 1976
From the Chargé d'affaires ad interim
Embassy of the United States of America
London

To the Minister of State for Foreign and Commonwealth Affairs
Foreign and Commonwealth Office
London

Appendices

Sir:

I have the honor to refer to the Diego Garcia Agreement 1976, constituted by the Exchange of Notes of today's date between the Government of the United Kingdom of Great Britain and Northern Ireland and the Government of the United States of America, supplementing the British Indian Ocean Territory Agreement (BIOT Agreement), effected by an Exchange of Notes between the two Governments dated December 30, 1966.

In accordance with the recent discussions between representatives of our two Governments, I have the honor to inform you that the Government of the United States of America, subject to the availability of funds, plans to undertake the following additional construction on Diego Garcia for the United Stales Navy support facility to be developed there:

Item	*Approximate Capacity or Size*
Expanded dredging for fleet anchorage	4,000 acres
Fuel and general purpose pier	550 feet of berthing
Runway extension	4,000 linear feet
Aircraft parking apron	90,000 square yards
Hangar	18,000 square feet
Air operating building addition	2,900 square feet
Transit storage building	4,000 square feet
Aircraft arresting gear	—
Storage petroleum, oil and lubricants	640,000 barrels
Power plant expansion	2,400 kilowatts
Vehicle repair hardstand	1,200 square yards
Subsistence building addition	3,600 square feet
Cold storage addition	4,200 square feet
Armed forces radio and television station	1,200 square feet
General warehouse addition	13,200 square feet
Utilities	—
Ready issue ammunition magazine	2,000 square feet
Protective open storage area for munitions	6,000 square yards
Bachelor enlisted quarters	277 men
Bachelor officers' quarters	32 men
Receiver building addition	1,300 square feet
Recreational facilities	(Scope to be determined)
Shed storage	7,100 square feet
Flammable storage	2,700 square feet
Navy exchange warehouse	5,400 square feet
Crash fire station	7,300 square feet
Structural fire station	3,000 square feet
Aircraft washrack	(Scope to be determined)
Aircraft ready issue refueler	(Scope to be determined)
Public works shops	16,600 square feet

The foregoing would be in addition to construction for the limited naval communications facility presently on Diego Garcia, regarding which information has previously been provided to United Kingdom authorities. In the event that further construction should be planned for the facility, it would, of course, be understood that such construction would be subject to the provisions of paragraph 2(b) of the BIOT Agreement as well as paragraph 1(a) or 1(b), as appropriate, of the Diego Garcia Agreement 1976.

Accept, Sir, the renewed assurances of my highest regard.

[signed: *Ronald I. Spiers*]

II

25 February 1976

From the Minister of State for
Foreign and Commonwealth Affairs
Foreign and Commonwealth Office
London

To the Chargé d'affaires ad interim
Embassy of the United States of America
London

Sir,

I have the honour to acknowledge receipt of your letter of today's date concerning your Government's plans for construction in connection with the development of the present limited naval communications facility on Diego Garcia as a support facility of the United States Navy. I note the additional construction planned by your Government and your statement concerning further construction. I confirm that your statement is in accordance with the understanding of my Government.

I have the honour to be with high consideration, Sir, Your obedient Servant,

[signed: *Roy Hattersley*]

V. Amendment London, June 1976[8]

Availability for Defence Purposes of the British Indian Ocean Territory

I

No. HKT 040/1 22 June 1976

From the Secretary of State for Foreign and Commonwealth Affairs
Foreign and Commonwealth Office, London

To the Ambassador of the United States of America
London

Your Excellency,

I have the honour to refer to recent discussions between representatives of the Government of the United Kingdom of Great Britain and Northern Ireland and the Government of the United States of America concerning the Agreement constituted by the Exchange of Notes dated 30 December 1966 concerning the availability of the British Indian Ocean Territory islands for defence purposes (hereinafter referred to as "the Agreement") and to propose that the Agreement be amended by deleting the following words in the opening paragraph: "and the islands of Aldabra, Farquhar and Desroches."

If the foregoing proposal is acceptable to the Government of the United States of America, I have the honour to propose that this Note and Your Excellency's reply to that effect shall constitute an Agreement between the two Governments to amend the Agreement of 30 December 1966 with effect from 28 June 1976.

I have the honour to be, with the highest consideration, Your Excellency's obedient Servant,

(For the Secretary of State)
[signed: *E.N. Larmour*]

II

From the Ambassador of the United States of America 25 June 1976
London

To the Secretary of State for Foreign and Commonwealth Affairs
Foreign and Commonwealth Office, London

Dear Secretary:

I have the honor to acknowledge receipt of your Note no. HKT 040/1 of June 22 which reads as follows:

[see Note *I* above]

In reply I have the honor to inform you that the foregoing proposal is acceptable to the Government of the United States of America which therefore approves Your Excellency's suggestion that your Note and this reply shall constitute an Agreement between the two Governments to amend the Agreement of 30 December 1966 with effect from 28 June 1976.

Accept, Sir, the assurances of my highest consideration.

[signed: *Anne Armstrong*]

VI. Supplement Washington, D.C., December 1982[9]

Further Supplementary Arrangements on Diego Garcia

T.I.A.S. 10616 December 13, 1982

Preamble

Pursuant to paragraph 19 of the February 25, 1976, Agreement effected by an exchange of notes between the Government of the United Kingdom of Great Britain and Northern Ireland and the Government of the United States of America concerning the United States Navy Support Facility on Diego Garcia, the Government of the United Kingdom of Great Britain and Northern Ireland represented by the Foreign and Commonwealth Office and the Government of the United States represented by the Department of State, desiring to make arrangements additional to those made in the Supplementary Arrangements signed on 25 February 1976, have made the following further supplementary arrangements which are to be read as one with the Supplementary Arrangements of 1976.

Paragraph 1: General

In determinating the number of its personnel stationed at any one time on the island and using the island's resources, the United States will give due consideration to the limited resources of the island, in particular drinking water, and the need to conserve the environment.

Paragraph 2: Contractor Personnel

If the United States Government intends to introduce large numbers of third country nationals, it will consult Her Majesty's Government, indicating in general terms the numbers concerned, the nationalities and the expected duration of their stay on Diego Garcia.

Paragraph 3: Fresh Water

From time to time, or if specially requested, the United States Commanding Officer will advise the British Government Representative of the adequacy of the fresh water supply on the island and of the arrangements for extracting drinking water. The British Government Representative will be informed of any significant fall in the level of the water table or any significant deterioration in the quality of the water, with a view to jointly agreed remedial action.

Paragraph 4: Dumping

There will be no dumping of vehicles, machinery, equipment or other non-natural waste in the territory of the Chagos Archipelago without the prior approval of the British Government Representative.

Paragraph 5: Dredging and Reef Blasting

It will be the responsibility of the United States Commanding Officer to ensure that no dredging or reef blasting is conducted in any area where it could cause irremediable damage. Normally, a breadth of live reef of no less than 80 yards will be left intact and free of blasting operations all around the island. If, exceptionally, it is considered necessary to reduce the breadth of the live reef to less than 80 yards at any point, the British Government Representative will be consulted in sufficient time in advance of the proposed action to enable an assessment of the likely ecological consequences to be made by qualified authorities.

Paragraph 6: Hold-Harmless Provision

Whenever, pursuant to paragraph 5 of the Agreement effected by an exchange of notes of December 30, 1966, the United States Government makes facilities available to the United Kingdom Government, the United Kingdom Government will indemnify and hold-harmless the United States Government for any liability incident to the use of such facilities by the United Kingdom Government, its agencies, agents, officers, employees, contractors or other users authorized by the United Kingdom Government.

December 13, 1982

[signed: *Nigel Wenban-Smith*]
On behalf of the Government
of the United Kingdom of
Great Britain
and Northern Ireland

[signed: *Jonathan T. Howe*]
On behalf of the Government
of the United States of
America

APPENDICES

VII. Supplement Washington, D.C., November 1987[10]

Operations and Construction Contracts on Diego Garcia

I

From the Secretary of State of
the United States of America
Department of State, Washington, D.C.

November 16, 1987

To Her Majesty's Ambassador to Washington
British Embassy, Washington, D.C.

Excellency:

I have the honor to refer to recent discussions between representatives of our two Governments regarding the exchange of notes between the Government of the United States of America and the United Kingdom of Great Britain and Northern Ireland concerning the availability for defense purposes of the British Indian Ocean Territory signed in London on December 30, 1966 (hereinafter referred to as the Agreement), and in particular paragraph 7 of that Agreement.

As a result of these discussions, and in the light of operational and security requirements, I have the honor to propose that the operations and construction contracts applicable to the United States military installations on Diego Garcia shall henceforth be awarded by United States military authorities to joint-ventures exclusively between United States and United Kingdom firms unless no qualified joint-venture firm submits a reasonable offer for such contracts, in which case they shall be opened to United States firms.

In addition, the following conditions shall apply to such contracts involving any United States-United Kingdom joint-venture:

(a) management control shall be vested in the United States partner or partners,
(b) the percentage of participation between United States and United Kingdom firms shall be a matter for negotiations between members of the joint-venture consortium, but in no event shall United States participation be less than sixty percent, or United Kingdom participation below twenty percent,

(c) a definition of a United States firm which may be more restrictive than that set forth in paragraph 10(e) of the Agreement shall be provided in each contract or solicitation.

If the foregoing proposal is acceptable to the Government of the United Kingdom of Great Britain and Northern Ireland, I have the honor to propose that the present note, together with Your Excellency's reply to that effect, shall constitute an agreement between the two Governments which shall enter into force on the date of Your Excellency's reply.

Accept, Excellency, the renewed assurances of my highest consideration.

For the Secretary of State:
[signed: *Harry Allen Holmes*]

II

From Her Majesty's Ambassador to Washington   16 November 1987
British Embassy, Washington, D.C.

To the Secretary of State of the United States of America
Department of State, Washington, D.C.

Your Excellency,
I have the honour to acknowledge receipt of your Excellency's note which reads as follows:

*[see Note **I** above]*

I have the honour to inform Your Excellency that the foregoing proposal is acceptable to the Government of the United Kingdom of Great Britain and Northern Ireland who therefore agree that Your Excellency's note and this reply shall constitute an agreement between the two Governments which shall enter into force on the date of this reply.

I take this opportunity to renew to Your Excellency the assurance of my highest consideration.

[signed: *Antony Acland*]

VIII. Agreement London, June/July 1999[11]

Construction of a Monitoring Facility on Diego Garcia, BIOT

I

OTI083/001/99
From the Head of Overseas Territories Department
Foreign and Commonwealth Office, London

18 June 1999

To the Ambassador of the United States of America
London

I have the honour to refer to the Agreement constituted by the Exchange of Notes dated 30 December 1966 between the Government of the United Kingdom of Great Britain and Northern Ireland and the Government of the United States of America concerning the availability of the British Indian Ocean Territory for defence purposes and to the Agreement constituted by the Exchange of Notes dated 25 February 1976 between the two Governments concerning a United States Navy Support Facility on Diego Garcia, British Indian Ocean Territory.

Pursuant to paragraph 2(b) of the former Agreement, I now convey the approval of the Government of the United Kingdom of Great Britain and Northern Ireland to the construction on Diego Garcia, British Indian Ocean Territory, of a monitoring facility for inclusion in the International Monitoring System to be established pursuant to the Comprehensive Nuclear Test-Ban Treaty, adopted at New York on 10 September 1996, and for satisfying requirements of the United States of America, and propose an Agreement in the following terms:

Scope of and responsibility for the monitoring facility

1. The Government of the United States of America shall undertake, at no expense to the Government of the United Kingdom of Great Britain and Northern Ireland, to construct, maintain and operate a hydro-acoustic monitoring facility on Diego Garcia, British Indian Ocean Territory, to meet US requirements and for inclusion in the International Monitoring System to be established, pursuant to the Comprehensive Nuclear Test-Ban Treaty, by the Preparatory Commission and its Provisional

Technical Secretariat. Notwithstanding that the monitoring facility shall be constructed, maintained and operated by the Government of the United States, the Government of the United Kingdom of Great Britain and Northern Ireland shall be the State responsible for the monitoring facility pursuant to Section A of Part 1 of the Protocol of the Comprehensive Nuclear Test-Ban Treaty.

Purpose of the monitoring facility

2. The purpose of the monitoring facility shall be to provide data to the International Data Centre and, as appropriate, the Provisional or Prototype International Data Centre, established by the Preparatory Commission and the Comprehensive Nuclear Test-Ban Treaty Organization in accordance with the terms of the Treaty; and directly to the United States of America. At its own expense, the Government of the United States of America may install and maintain a separate data channel for its own purposes.

Compliance with the Comprehensive Nuclear Test Ban Treaty

3. The monitoring facility shall be surveyed, constructed, operated and maintained in accordance with US requirements and applicable provisions of the Comprehensive Nuclear Test-Ban Treaty, the technical specifications approved by the Preparatory Commission on 18 August 1998 and the associated Operational Manuals adopted by the Preparatory Commission of the Comprehensive Nuclear Test-Ban Treaty Organization. The two Governments shall take any measures necessary to work, as appropriate, with the Preparatory Commission or the Comprehensive Nuclear Test-Ban Treaty Organization to ensure that the station will be certified to operate as an International Monitoring System [IMS] station. Given the obligation of the Organization in Article IV of the Treaty and the provisional obligation of the Preparatory Commission in paragraph 5(c) of the Annex to the Resolution Establishing the Preparatory Commission for the Comprehensive Nuclear Test-Ban Treaty to meet the costs of IMS facilities, the US and the UK shall cooperate in any effort to obtain funding from the Organization or the Preparatory Commission, as appropriate, for some or all of the costs of establishing, operating, and maintaining this station. Installation, operation, maintenance, and funding of any future upgrades of the station following initial certification will be a matter for consultation and agreement between Governments.

APPENDICES 113

4. The Governments shall cooperate with the Comprehensive Nuclear Test-Ban Treaty Organization to provide a direct connection from the monitoring facility to the Global Communications Infrastructure (GCI) of the Treaty Organization to ensure compliance with the Comprehensive Nuclear Test-Ban Treaty and its Preparatory Commission requirements. Continuous data from the monitoring facility will be transmitted using the GCI, uninterrupted to the International Data Centre and, as appropriate, the Provisional or Prototype International Data Centre.

Respect for British Indian Ocean Territory laws and regulations

5. The Government of the United States shall, with regard to the construction and operation of the monitoring facility, respect all applicable British Indian Ocean Territory laws and regulations, and shall, as far as possible, minimise any adverse impact that the monitoring facility might have on the local environment.

Construction of the monitoring facility

6. The plans for the construction of the monitoring facility, and its precise location, shall be agreed between the two Governments, and they shall jointly review the construction plans with the Provisional Technical Secretariat of the Comprehensive Nuclear Test-Ban Treaty Preparatory Commission prior to commencement of construction. Prior to, and during, construction of the monitoring facility, the Government of the United States shall keep the Government of the United Kingdom of Great Britain and Northern Ireland informed of progress and of any significant difficulties encountered or envisaged.

Oversight

7. The Government of the United Kingdom of Great Britain and Northern Ireland shall have the right to oversee planning, construction, maintenance and operation of the monitoring facility and, given its overall responsibility for the facility, will serve as the conduit of information on this facility to the Preparatory Commission of the Comprehensive Nuclear Test-Ban Treaty Organization.

Reduced Assessment

8. As set forth in Article IV, paragraph 22 of the Comprehensive Nuclear Test-Ban Treaty, and the appropriate provisions of the

Financial Regulations of the Preparatory Commission for the Comprehensive Nuclear Test-Ban Treaty Organization, decisions of the Preparatory Commission (for example, CTBT/PC/III/CRP.2/Rev.2 and CTBT/PC/II/CRP.15/Rev. 1) and taking into account any subsequent decisions taken by the Preparatory Commission and the Comprehensive Nuclear Test-Ban Treaty Organization, the two Governments may submit requests, either independently or jointly, for a reduced assessment to the Preparatory Commission or the Executive Council. Prior to such request, the two Governments shall have agreed on the division of any funding that each Government shall have contributed toward the establishment/upgrade of the station, and the Government of the United Kingdom of Great Britain and Northern Ireland shall have informed the Comprehensive Nuclear Test-Ban Treaty Preparatory Commission or the Executive Council of the results of such agreements.

Notification

9. The Government of the United Kingdom of Great Britain and Northern Ireland shall register this Agreement with the Secretary General of the United Nations and shall inform the Provisional Technical Secretariat of the Comprehensive Nuclear Test-Ban Treaty Organization of its conclusion.

Availability of funds

10. To the extent that the carrying out of any activity or the implementation of any part of this Agreement depends upon funds to be appropriated by the Congress of the United States, it shall be subject to the availability of such funds.

Duration

11. This Agreement shall continue in force for as long as the British Indian Ocean Territory Agreement of 30 December 1966 continues in force, or until such time as the parties agree that no part of Diego Garcia is any longer required for the purpose of the facility as set forth in paragraph 2, whichever occurs first. After this Agreement has been in force for ten years, either party may terminate it upon one-year notice.

Options for future infrasound and radionuclide stations

12. The Governments of the United States of America or the United Kingdom of Great Britain and Northern Ireland may,

subject to the availability of funds and mutual agreement between the two Governments, later include additional monitoring facilities consisting of an infrasound station and a radionuclide station.

If the Government of the United States of America is content with these proposals, I have the honour to propose that this Note together with your reply to that effect shall constitute an Agreement between our two Governments which shall enter into force on the date of your reply and shall be known as the British Indian Ocean Territory Agreement 1999.

I have the honour to convey to Your Excellency the assurance of my highest consideration.

[signed: *C J B White*]

II

No. 045   21 July 1999
From the Ambassador of the United States of America
London

To H.E. C.J.B. White, Head of the Overseas Territories Department Foreign and Commonwealth Office, London

Your Excellency:

I have the honor to acknowledge receipt of your letter number *OTI083/001/99* of 18 June 1999 which reads as follows:

*[see Note **I** above]*

In reply I have the honor to inform you that the foregoing proposal is acceptable to the Government of the United States of America which therefore approves your Excellency's uggestion that your note and this reply shall constitute an Agreement between the two Governments which shall enter into force on the date of this note.

Accept, Excellency, the renewed assurances of my highest consideration.

[signed: *Philip Lader*]

IX. Exchanges of Letters, London–Washington, D.C., 2001–2004[12]

1. Infrastructure Upgrades, Bomber Forward Operating Location Diego Garcia, August/September 2001

In a letter dated August 28, 2001, from the U.S. ambassador to London (William S. Farish III) to the BIOT commissioner (Alan E. Huckle, head of overseas territories in the Foreign and Commonwealth Office), the U.S. Government proposed that the U.S. Naval Support Facility at Diego Garcia, British Indian Ocean Territory, be designated as a "Bomber Forward Operating Location" (FOL), "such designation being consistent with the scope and purpose of the facility pursuant to paragraphs (1) and (2) of the Diego Garcia Agreement 1976."

In his response letter dated September 10, 2001, the BIOT commissioner confirmed that the arrangements so proposed (see the annex below) were acceptable to the UK government.

Annex

The Government of the United States of America, subject to the availability of funds, plans to undertake the following infrastructure upgrades:

Air Expeditionary Force ("AEF") Infrastructure Upgrades
Pre-Engineered Buildings (PEBs)—USD $5m
—Construction of 100 units on existing concrete pads
—Construction of three small masonry buildings to house the power distribution units that will provide electricity to the PEBs
—Construction of a small concrete pad for an emergency back-up power generator
—Paving of access roads within the tent area
Munitions Storage Magazine—USD $5.5m
—Conversion of four (4) open air bunkers to modular munitions storage magazines
—Current clear "safety" zone requirements would still apply
Land Mobile Radar Repeater—USD $400k

—Construction of a (7x7 foot) building adjacent to the water tower to shelter support Equipment
Upgrade of Existing Classified Storage Facilities—USD $400k
—At the present time, no further planned construction is anticipated in support of the AEF upgrade. In the event that further construction should be required, the U.S. Government will consult with the UK Government in accordance with our agreements and understandings and past practice.

U.S. Naval Facility Infrastructure Upgrades
(not specifically related to FOL designation)
Aircraft Intermediate Maintenance Facility—USD $8.15m
Physical Fitness Facility—USD (to be determined)
—Details of construction costs have not yet been provided. This project is proposed as a quality of life issue for personnel serving on Diego Garcia.

2. Additional Infrastructure Upgrades on Diego Garcia, July 2002

In a letter dated July 23, 2002, from the director, Office of International Security Operations in the U.S. State Department's Bureau of Political-Military Affairs (Col. Charles Wilson), to the BIOT commissioner (Alan E. Huckle, head of overseas territories in the Foreign and Commonwealth Office), the U.S. government—referring to the 1966 and 1976 agreements, and to the "annual U.S.-U.K. Diego Garcia talks held on May 14, 2002"—proposed additional infrastructure upgrades to the U.S. Naval Support Facility at Diego Garcia, British Indian Ocean Territory.

In his response letter dated July 29, 2002, the BIOT commissioner confirmed that the arrangements so proposed (see the annex below) were acceptable to the UK government.

Annex: US Construction Proposals

—*B2 Shelters (FOL Agreement)*
The number one issue for the US; the ability to station B2s at Diego Garcia would reduce the current 50-hour round trip from Missouri to something in the order of 15 hours.
—*Improved data telephone communications system (FOL Agreement)*
—*Construction of an ammunition handling facility (FOL Agreement)*

—*Air Mobility Command Squadron Operations building* (akin to an airport terminal)
—*Enhancement of existing temporary living encampments (FOL Agreement)*
—*Transient berthing and medical facility* (temporary accommodation for personnel transiting DG)
—*Enhancements to the aircraft fuelling system (FOL Agreement)*
—*Runway improvements*
—*Air defence radar system*
—*Air Traffic Control radar system.*

3. Satellite Tracking Station on Diego Garcia, November/December 2004

In a letter dated November 29, 2004, from the assistant secretary of state for political-military affairs in the U.S. State Department (Lincoln P. Bloomfield Jr.) to the BIOT commissioner (Robert N. Culshaw MVO, director of the Americas and overseas territories in the Foreign and Commonwealth Office), the U.S. government—referring to the "U.S.-UK Pol-Mil Talks 2004"—proposed to formalize the UK government's (a) approval of the existing satellite tracking station, and (b) approval of a U.S. request to expand the satellite tracking station "to provide enhanced capability and more flexibility in conducting maintenance."

In his response letter dated December 7, 2004, the BIOT commissioner confirmed that the arrangements so proposed (see the annex below) were acceptable to the UK government.

Annex: U.S. Request for Expansion of the Satellite Tracking Station at Diego Garcia

1. Foreign Operating Rights:
   a. *Project name*
      Remote Tracking Station Block Change (RBC) Antenna radome, and associated core electronic/support equipment at Diego Garcia BIOT.
   b. *Description of project*
      The proposed project would install an RBC system at Diego Garcia BIOT. RBC consists of the following major components:
      (1) 13-meter dish antenna for satellite control, with an inflatable radome and support equipment, and ops core electronic equipment. The radome is the outward appearance of this

system, protects the antenna from the weather, and is virtually identical to what was installed at TCS Oakhanger [RAF station in Hampshire/UK] in 2000 to replace an obsolete antenna [photo omitted].

(2) The project provisions for deployment of a transportable asset to serve in times of surge, contingency, and/or other operational needs. As such, the addition of a transportable pad and associated electrical and communication interfaces to/from the pad will be assessed and made as appropriate.

c. *USAF requirements*

USAF requires additional capability to control satellites from Diego Garcia. Although there is one Air Force Satellite Control Network (AFSCN) antenna already operating at Diego Garcia BIOT, USAF requirements for telemetry reception and commanding of satellites supporting military missions in areas of mutual strategic interest have reached the point where an additional antenna is required.

d. *Method of operation*

(1) The antenna will operate using the same methods, and in the same frequency range and power levels as the current AFSCN antenna operating at Diego Garcia. The antenna will be active repeatedly, but not continuously throughout its 24 hour operating cycle. Antenna transmit/receive operations will be scheduled at varying power levels up to a maximum of 2000 Watts.

(2) to ensure safety of operating personnel as well as those living in the surrounding area, the new system is being designed to strict safety guidelines. USAF will transmit commands and receive telemetry to/from satellites through control centers located at Schriever Air Force Base (AFB), Colorado, and Onizuka Air Force Station, California. Similar to today's operation, a daily operations schedule will be provided by Schriever AFB to the AFSCN site located at Diego Garcia BIOT. For each scheduled satellite contact the following will occur:

(a) A Satellite Operations Center (SOC) in the US will be connected to Diego Garcia BIOT through communications links

(b) Equipment will be configured and checked out

(c) Commands will be sent from the SOC via Diego Garcia BIOT through the new antenna and up to a satellite

(d) The satellite will send back telemetry data (such as health and status of onboard communications, thermal control or attitude control systems) through the new Diego Garcia BIOT antenna, which will be routed back to the SOC for processing

(e) At completion of the scheduled events, there will be an "idle period" until the next scheduled operation when the sequence described above will be repeated.

(3) Equipment: A suite of approximately seven (7) racks of electronic equipment, located inside the existing operations building, would perform equipment configuration, control, monitoring, signal conditioning and routing functions. The equipment would be introduced nt the site by commercial transportation and installed by USAF's contractor who is developing the system. When the system is turned over for operations, the equipment will be USAF owned. Additional communications capacity will also be added.

(4) AFSPC on-island contractor personnel would perform operations and normal "remove and replace" maintenance. USAF personnel, as required, would provide Level 2 "depot-level" maintenance.

(5) The proposed location for the new antenna is between the Diego Garcia current "A" antenna and the GPS antenna near the current cement transportable pad, so as to use existing site real estate, but nonetheless outside the current Diego security fence. This would require an area to reclaim lands outside the existing fence line for security clear zones.

e. *Personnel involved*

Current Diego Garcia BIOT staffing consists of an Air Force officer in command with a contractor support staff, which performs all operations, "remove and replace" maintenance, logistics, and support functions. Additional permanent USAF personnel are required at Diego Garcia BIOT for this new antenna, currently projected at six individuals to cover shift responsibilities. When the automated features of the system are fully evaluated and realized, some future reduction in that number of personnel is possible.

f. *Special requirements*

There are no special requirements for billeting, mess, or transportation for this project. A request for Radio Frequency Authorization will be processed through the normal channels for approval.

g. *Classification of project*

Unclassified, releasable to UK.

h. *Release of project data to host nation*

Cognizant authority is SMC/AXP, Los Angeles Air Force Base, California. This office is a designated foreign disclosure authority. Project planning data required by UKMOD [United Kingdom Ministry of Defence] to act upon this request will be provided through SMC/RNA, Los Angeles Air Force Base, California. Technical data will be handled through normal export control, review, and release procedures. Data required for UKMOD Diego Garcia BIOT personnel to operate and maintain the proposed RBC system will be provided through existing procedures.

i. *Possible participation or observation by local officials*

USAF encourages participation by UK representatives. Decisions on local officials' participation or observation will be at the discretion of UKMOD. No aerial flight operations are expected as part of this project.

j. *Potential benefit of project to the host nation*

The introduction of the proposed RBC system further strengthens the USAF-UKMOD relationship and provides an avenue of cooperation which has proven very beneficial to support special, emergency UKMOD requests during past armed conflicts. An example of this is when UKMOD requested urgent access to weather data in areas not served by commercial systems. USAF responded immediately to meet UKMOD military operational needs.

k. *Timeline to IOC*

DGS critical design review: June 2005
DGS install period: May 2006– June 2007
DGS test period: July 2007– November 2007
DGS operational date: January 2008.

X. House of Lords Judgment, London, October 2008[13]

The Queen (on the application of Bancoult) vs. Secretary of State for Foreign and Commonwealth Affairs

Opinion of Lord Hoffmann (for the majority)

1. This appeal concerns the validity of section 9 of the British Indian Ocean Territory (Constitution) Order 2004 ("the Constitution Order"):
   (1) Whereas the Territory was constituted and is set aside to be available for the defence purposes of the Government of the United Kingdom and the Government of the United States of America, no person has the right of abode in the Territory.
   (2) Accordingly, no person is entitled to enter or be present in the Territory except as authorised by or under this Order or any other law for the time being in force in the Territory.
2. The Constitution was made by prerogative Order in Council. The Divisional Court (Hooper LJ and Cresswell J) held section 9 to be invalid and this decision was affirmed by the Court of Appeal (Sir Anthony Clarke MR and Waller and Sedley LJJ). The Secretary of State appeals to your Lordships' House.
3. The British Indian Ocean Territory ("BIOT") is situated south of the equator, about 2200 miles east of the coast of Africa and 1000 miles south-west of the southern tip of India. It consists of a group of coral atolls known as the Chagos Archipelago of which the largest, Diego Garcia, has a land area of about 30 km^2. Some distance to the north lie Peros Banhos (13 km^2) and the Salomon Islands (5 km^2).
4. The islands were a dependency of Mauritius when it was ceded to the United Kingdom by France in 1814 and until 1965 were administered as part of that colony. Their main economic activity was gathering coconuts and extracting and selling the copra or kernels. In 1962, when the plantations were acquired by a Seychelles company called Chagos Agalega Company Ltd ("the company") the settled population was a very small community (less than 1,000 on the three islands) who called themselves *Ilôis* (*Créoles des Îles*) and whose families had in some cases lived in the islands for generations. With the assistance of contract labour from the Seychelles and Mauritius, the *Ilôis* were

mainly employed in tending the coconut trees and producing the copra.

5. The evidence suggests that the Îlois, who now prefer to be called Chagossians, lived an extremely simple life. The company, whose managers acted as justices of the peace, ran the islands in feudal style. Each family had a house with a garden and some land to provide vegetables, poultry and pigs to supplement the imported provisions supplied by the company. They also did some fishing. There was work in the copra industry as well as some construction, boat building and domestic service for the women. No one was involuntarily unemployed. Most of the Chagossians were illiterate and their skills were confined to those needed for the activities on the islands. But they had a rich community life, the Roman Catholic religion and their own distinctive dialect derived (like those of Mauritius and the Seychelles) from the French.

6. Into this innocent world there intruded, in the 1960s, the brutal realities of global politics. In the aftermath of the Cuban missile crisis and the early stages of the Vietnam War, the United States felt vulnerable without a land based military presence in the Indian Ocean. A survey of available sites suggested that Diego Garcia would be the most suitable. In 1964 it entered into discussions with Her Majesty's Government which agreed to provide the island for use as a base. At that time the independence of Mauritius and the Seychelles was foreseeable and the United States was unwilling that sovereignty over Diego Garcia should pass into the hands of an independent "non-aligned" government. The United Kingdom therefore made the British Indian Ocean Territories Order 1965 SI No 1920 ("the BIOT order") which, under powers contained in the Colonial Boundaries Act 1895, detached the Chagos Archipelago (and some other islands) from the colony of Mauritius and constituted them a separate colony known as BIOT. The order created the office of Commissioner of BIOT and conferred upon him power to "make laws for the peace, order and good government of the Territory." Those inhabitants of BIOT who had been citizens of the United Kingdom and Colonies by virtue of their birth or connection with the islands when they were part of Mauritius retained their citizenship. When Mauritius became independent in 1968 they acquired Mauritian citizenship but, by an exception in the Mauritius Independence Act 1968, did not lose their UK citizenship.

7. At the end of 1966 there was an exchange of notes between Her Majesty's Government and the Government of the United States by which the United Kingdom agreed in principle to make BIOT available to the United States for defence purposes for an indefinitely long period of at least 50 years. It subsequently agreed to the establishment of the base on Diego Garcia and to allow the United States to occupy the other islands of the Archipelago if they should wish to do so.
8. In 1967 the United Kingdom Government bought all the land in the Archipelago from the company but granted the company a lease to enable it to continue to run the coconut plantations until the United States needed vacant possession. It took some time for the US Defence Department to obtain Congressional approval but in 1970 it gave notice that Diego Garcia would be required in July 1971. After receiving this notice the Commissioner of BIOT, using his powers of legislation under the BIOT order, made the Immigration Ordinance 1971. It provided in section 4(1) that—
   no person shall enter the Territory or, being in the Territory, shall be present or remain in the Territory, unless he is in possession of a permit... [issued by an Immigration Officer]
9. Between 1968 and 1971 the United Kingdom government secured the removal of the population of Diego Garcia, mostly to Mauritius and the Seychelles. A small population remained on Peros Banhos and the Salomon Islands, but they were evacuated by the middle of 1973. No force was used but the islanders were told that the company was closing down its activities and that unless they accepted transportation elsewhere, they would be left without supplies. The whole sad story is recounted in detail in an appendix to the judgment of Ouseley J in *Chagos Islanders v Attorney General* [2003] EWHC 2222 (QB), [2003] All ER (D) 166.
10. My Lords, it is accepted by the Secretary of State that the removal and resettlement of the Chagossians was accomplished with a callous disregard of their interests. For the most part, the community was left to fend for itself in the slums of Port Louis. The reasons were to some extent the usual combination of bureaucracy and Treasury parsimony but very largely the government's refusal to acknowledge that there was any indigenous population for which the United Kingdom had a responsibility. The Immigration Ordinance, denying that anyone was entitled to enter or live in the islands, was part of the legal façade constructed to defend this claim. The government adopted this

position because of a fear (which may well have been justified) that the Soviet Union and its "non-aligned" supporters would use the Chagossians and the United Kingdom's obligations to the people of a non-self-governing territory under article 73 of the United Nations Charter to prevent the construction of a military base in the Indian Ocean.

11. When the Chagossians arrived in Mauritius they found themselves in a country with high unemployment and considerable poverty. Their conditions were miserable. There was a long period of negotiation between the governments of Mauritius and the United Kingdom over payment for the cost of resettlement, but eventually in September 1972 the two governments agreed on a payment of £650,000, which was paid in March 1973. The Mauritius government did nothing with the money until 1977 when, depleted by inflation, it was distributed in cash to 595 Chagossian families.

12. The Chagossians sought support and legal advice. In February 1975 Michael Vencatessen, who had left Diego Garcia in 1971, issued a writ in the High Court in London against the Foreign and Defence Secretaries and the Attorney General. His proceedings were funded by legal aid and he received the advice of distinguished counsel. The claim was for damages for intimidation and deprivation of liberty in connection with his departure from Diego Garcia, but the proceedings came to be accepted on both sides as raising the whole question of the legality of the removal of the Chagossians from the islands.

13. Negotiations took place between the UK government and Mr Vencatessen and his advisers, who were treated as acting on behalf of the Chagossians as a whole. In 1979 an agreement was reached with Mr Vencatessen and his advisers for a payment of £1.25m in settlement of all the claims of the Chagossians. His solicitor went to Mauritius to seek the approval of the community but was unable to obtain it. Further negotiations, in which the government of Mauritius participated, took place over the next three years. Finally in July 1982 it was agreed that the UK government would pay £4m into a trust fund for the Chagossians, set up under a Mauritian statute. The agreement was signed by the two governments in the presence of Chagossian representatives and provided for individual beneficiaries to sign forms renouncing all their claims arising out of their removal from the islands. About 1340 did so, but a few did not.

14. At that point the UK government might have thought that, however badly its predecessors in office may have behaved in securing the removal of the Chagossians from the islands, the matter was now settled and a line could be drawn under this unfortunate episode. Any such hope would have been disappointed. Sixteen years later, on 30 September 1998 Mr Bancoult, the applicant in these proceedings, applied for judicial review of the Immigration Ordinance 1971 and a declaration that it was void because it purported to authorise the banishment of British Dependent Territory citizens from the Territory and a declaration that the policy which prevented him from returning to and residing in the Territory was unlawful.

15. The government's reaction to the institution of these proceedings was to commission an independent feasibility study to examine whether it would be possible to resettle some of the Chagossians on the outer islands of Peros Banhos and the Salomon Islands. There was no question of their return to Diego Garcia, which the United States was entitled to occupy until at least 2016. It must have been clear to both parties that the challenge to the validity of the 1971 Ordinance was largely symbolic. There was no evidence that it had ever been used to expel anyone from the islands. The islanders who left between the time it was made and the final evacuation in 1973 did so because they were left with the alternative of being abandoned without support or supplies. Nor would its revocation have any practical effect on whether the Chagossians could go back and reside there. That would require an investment in infrastructure and employment which the Chagossians could not themselves provide. As was demonstrated by subsequent actions, the judicial review proceedings were only a part of a new campaign by the Chagossians to obtain UK government support for their resettlement to right the wrongs of 1968–1973.

16. On 3 November 2000 the Divisional Court (Laws LJ and Gibbs J) gave judgment in favour of Mr Bancoult: see *R (Bancoult) v Secretary of State for Foreign and Commonwealth Affairs* [2001] QB 1067 ("*Bancoult (1)*") They decided that a power to legislate for the "peace, order and good government" of the Territory did not include a power to expel all the inhabitants. The relief granted was an order quashing section 4 of the Immigration Ordinance as ultra vires.

17. After the judgment had been given, the Foreign Secretary (Mr Robin Cook) issued a press release:

Following the judgment in the BIOT Case on 3 November, Foreign Secretary Robin Cook issued the following statement:

I have decided to accept the Court's ruling and the Government will not be appealing.

The work we are doing on the feasibility of resettling the Ilois now takes on a new importance. We started the feasibility work a year ago and are now well underway with phase two of the study.

Furthermore, we will put in place a new Immigration Ordinance which will allow the Ilois to return to the outer islands while observing our Treaty obligations.

This Government has not defended what was done or said thirty years ago. As Lord Justice Laws recognised, we made no attempt to conceal the gravity of what happened. I am pleased that he has commended the wholly admirable conduct in disclosing material to the Court and praised the openness of today's Foreign Office.

18. On the same day, the Commissioner revoked the 1971 Immigration Ordinance and made the Immigration Ordinance 2000. This largely repeated the provisions of the previous Ordinance but contained a new section 4(3) which provided that the restrictions on entry or residence imposed by section 4(1) should (with the exception of Diego Garcia) not apply to anyone who was a British Dependent Territories citizen by virtue of his connection with BIOT.

19. As was to be expected, the change in the law made no practical difference. Some Chagossians made visits to the outer islands to tend family graves or simply to see and try to recognise their former homeland, but such visits had been made by permit under the old Ordinance and were invariably funded by the BIOT. No one went to live there. They awaited the report of the feasibility study.

20. In April 2002, before the production of the report, a group action was commenced on behalf of the Chagos Islanders against the Attorney General and other ministers, claiming compensation and restoration of the property rights of the islanders and declarations of their entitlement to return to all the Chagos Islands and to measures facilitating their return. On 9 October 2003 Ouseley J struck out this action on the grounds that the claim to

more compensation after the settlement of the Vencatessen case was an abuse of process, that the facts did not disclose any arguable causes of action in private law and that the claims were in any case statute-barred.

21. The importance of this judgment was that it unequivocally affirmed the validity of the 1982 settlement. The UK government had discharged its obligations to the Chagossians by payment in full and final settlement.

22. On 22 July 2004 the Court of Appeal (Dame Elizabeth Butler-Sloss P, Sedley and Neuberger LJJ) refused leave to appeal. Sedley LJ, who gave the judgment of the court, ended by saying:

> This judgment brings to an end the quest of the displaced inhabitants of the Chagos Islands and their descendants for legal redress against the state directly responsible for expelling them from their homeland. They have not gone without compensation, but what they have received has done little to repair the wrecking of their families and communities, to restore their self-respect or to make amends for the underhand official conduct now publicly revealed by the documentary record. Their claim in this action has been not only for damages but for declarations securing their right to return. The causes of action, however, are geared to the recovery of damages, and no separate claims to declaratory relief have been developed before us. It may not be too late to make return possible, but such an outcome is a function of economic resources and political will, not of adjudication.

23. The question of economic resources was of course what the feasibility study had been commissioned to investigate. The report was produced in June 2002. It concluded that "agroforestal production would be unsuitable for commercial ventures." So there could be no return to gathering coconuts and selling copra. Fisheries and mariculture offered opportunities although they would require investment. Tourism could be encouraged, although there was nowhere that aircraft could land. It might on be feasible in the short term to resettle the islands, although the water resources were adequate only for domestic rather than agricultural or commercial use. But looming over the whole debate was the effect of global warming which was raising the sea level and already eroding the corals of the low lying atolls. In the long term, the need for

sea defences and the like would make the cost of inhabitation prohibitive. On any view, the idyll of the old life on the islands appeared to be beyond recall. Even in the short term, the activities of the islanders would have to be very different from what they had been.

24. There followed discussion of the report between the Government (represented by Baroness Amos, Parliamentary Under-Secretary of State at the Foreign Office) and the applicant Mr Bancoult, his advisers and other representatives of the Chagossians. The Government was unwilling to commit itself one way or the other to a definite policy on resettlement until the Chagos Islanders action, which was claiming a legal entitlement to resettlement, had been resolved. But it resisted attempts on the part of the islanders to claim that the Foreign Secretary's press announcement and the revocation of the 1971 Immigration Ordinance amounted already to the adoption of a policy of resettlement. That decision would have to await the outcome of the litigation.

25. The judgment of Ouseley J in October 2003 made it clear that there was no legal obligation upon the United Kingdom, whether by way of additional compensation or otherwise, to fund resettlement. The government did not make any immediate statement, presumably because until 22 July 2004 there was still the possibility of an appeal. Before then, however, there was a development which gave the government concern. Newspaper articles appeared in Mauritius suggesting that the Chagossians and their supporters (principally a political group in Mauritius calling itself LALIT) were planning some form of direct action by landings on the islands. A '*flotille de la paix*' would be assembled to take some of the Chagossians to Diego Garcia or the outer islands. As might be expected, the various participants in this project had somewhat different aims. For LALIT, it was part of an anti-American campaign to close the base at Diego Garcia. Mr Bancoult did not want the base closed (he hoped it might employ resettled Chagossians) but was willing to lead a landing on the outer islands. In either case, since permanent resettlement on the islands was impractical without substantial investment, the landings, even if followed by temporary camps, could be no more than gestures in furtherance of the respective political aims of the parties, designed to attract publicity and embarrass the governments of the United Kingdom and the

United States. (On 12 March 2008 the *Guardian* reported that two British "human rights campaigners" had been arrested by the off Diego Garcia. They said that they were part of a group called the People's Navy which has been seeking to highlight the plight of the Chagossians and to protest against the military use of the islands.)

26. The Foreign Office was advised by its High Commission in Mauritius that the possibility of landings on the islands in the autumn of 2004 should be taken seriously. The United States also informed the UK government of its concern at any action which might compromise what it regarded as the unique security of Diego Garcia. The government had decided that in view of the feasibility report, it would not support resettlement of the islands. It therefore decided to restore full immigration control. On 10 June 2004 Her Majesty made the Constitution Order which revoked the BIOT Order and granted a new constitution including section 9, which I quoted at the commencement of my speech. At the same time, another Order in Council (the Immigration Order") was made dealing with the details of immigration control. In a written statement to the House of Commons on 15 June 2004 the Foreign Office Under Secretary of State Mr Bill Rammell explained that in the light of the feasibility report it would be "impossible for the Government to promote or even permit resettlement to take place. After long and careful consideration, we have therefore decided to legislate to prevent it."

27. The Minister went on to say that there had been "developments in the international security climate" since the judgment in *Bancoult (1)*" to which "due weight has had to be given." He did not mention the threatened landings which precipitated the decision to legislate, but the Foreign Secretary, in a letter dated 9 July 2004 to the chairman of the Foreign Affairs Committee of the House of Commons explaining why the Committee had not been shown the Constitution Order in draft before it was made, said that "we needed to preserve complete confidentiality if we were to avoid the risk of an attempt by the Chagossians to circumvent the Orders before they came into force."

28. These proceedings were commenced by an application for judicial review dated 24 August 2004, applying for section 9 of the Constitution Order and the Immigration Order to be quashed. The Divisional Court (paras 120–122) accepted an argument

that the Orders were irrational because their rationality had to be judged by the interests of BIOT. That meant the people who lived or used to live on BIOT. The Orders were not made in the interests of the Chagossians but in the interests of the United Kingdom and the United States and were therefore irrational.

29. This reasoning was not adopted, at any rate in quite the same form, by the Court of Appeal. Sedley LJ came nearest when he said that the removal or subsequent exclusion of the population "for reasons unconnected with their collective wellbeing" could not be a legitimate purpose of the power of colonial governance exercisable by Her Majesty in Council. It was an abuse of that power. He also considered that the Foreign Secretary's press statement after the judgment in *Bancoult (1)* and the Immigration Ordinance 2000 were promises to the Chagossians which gave rise to a legitimate expectation that, in the absence of a relevant change of circumstances, their rights of entry and abode in the islands would not be revoked. There had been no such change.

30. The Master of the Rolls and Waller LJ agreed that the applicant was entitled to succeed on the ground of a legitimate expectation. The Master of the Rolls also agreed with Sedley LJ that the Orders were an abuse of power because (see para 123) "they did not have proper regard for the interests of the Chagossians."

31. Before your Lordships the case has been most ably argued by Mr Jonathan Crow QC for the Crown and Sir Sydney Kentridge QC for the respondent. It is common ground that as BIOT was originally ceded to the Crown, Her Majesty in Council has plenary power to legislate for the Territory. The law is stated in Halsbury's *Laws of England* (4th ed 2003 reissue) vol 6, para 823:

> In a conquered or ceded colony the Crown, by virtue of its prerogative, has full power to establish such executive, legislative, and judicial arrangements as this Crown thinks fit, and generally to act both executively and legislatively, provided the provisions made by the Crown do not contravene any Act of Parliament extending to the colony or to all British possessions. The Crown's legislative and constituent powers are exercisable by Order in Council, Letters Patent or Proclamation.

32. Authority for these propositions will be found in Lord Mansfield's judgment in *Campbell v Hall* (1774) 1 Cowp 204 ("no question was ever started before, but that the King has a right to a legislative authority over a conquered country.") This

appeal requires your Lordships to determine the limits of that power.

33. On this point, both sides put forward what I would regard as extreme propositions. On the one hand, Mr Crow argued the courts had no power to review the validity of an Order in Council legislating for a colony. This was either because it was primary legislation having unquestionable validity comparable with that of an Act of Parliament, or because review was excluded by the terms of the Colonial Laws Validity Act 1865. On the other hand, Sir Sydney submitted that a right of abode was so sacred and fundamental that the Crown could not in any circumstances have power to remove it. Only an Act of Parliament could do so. I would reject both of these propositions.

34. It is true that a prerogative Order in Council is primary legislation in the sense that the legislative power of the Crown is original and not subordinate. It is classified as primary legislation for the purposes of the Human Rights Act 1998: see paragraph (f)(i) of the definition in section 21(1). That means that it cannot be overridden by Convention rights. The court can only make a declaration of incompatibility under section 4.

35. But the fact that such Orders in Council in certain important respects resemble Acts of Parliament does not mean that they share all their characteristics. The principle of the sovereignty of Parliament, as it has been developed by the courts over the past 350 years, is founded upon the unique authority Parliament derives from its representative character. An exercise of the prerogative lacks this quality; although it may be legislative in character, it is still an exercise of power by the executive alone. Until the decision of this House in *Council of Civil Service Unions v Minister for the Civil Service* [1985] AC 374, it may have been assumed that the exercise of prerogative powers was, as such, immune from judicial review. That objection being removed, I see no reason why prerogative legislation should not be subject to review on ordinary principles of legality, rationality and procedural impropriety in the same way as any other executive action. Mr Crow rightly pointed out that the *Council of Civil Service Unions* case was not concerned with the validity of a prerogative order but with an executive decision made pursuant to powers conferred by such an order. That is a ground upon which, if your Lordships were inclined to distinguish the case, it would be open to you to do so. But I see no reason for making

such a distinction. On 21 February 2008 the Foreign Secretary told the House of Commons that, contrary to previous assurances, Diego Garcia had been used as a base for two extraordinary rendition flights in 2002 (Hansard (HC Debates), cols 547–548). There are allegations, which the US authorities have denied, that Diego Garcia or a ship in the waters around it have been used as a prison in which suspects have been tortured. The idea that such conduct on British territory, touching the honour of the United Kingdom, could be legitimated by executive fiat, is not something which I would find acceptable.

36. The argument based on the Colonial Laws Validity Act is rather more arcane. The background to the Act is the statement of Lord Mansfield in *Campbell v Hall* (1774) 1 Cowp 204, 209 that although the King had power to introduce new laws into a conquered country, he could not make "any new change contrary to fundamental principles." If the King's power did not extend to making laws contrary to fundamental principles (presumably, of English law) in conquered colonies, it was regarded as arguable, in the first half of the nineteenth century, that the same limitation applied to the legislatures of settled colonies. It was never altogether clear what counted as fundamental principles and the Colonial Laws Validity Act was intended to put the question to rest by providing that no colonial laws should be invalid by reason of repugnancy to any rule of English law except a statute extending to the colony. In section 1 it defined "colonial law" as a law made for a colony by its legislature or by Order in Council. It defined "colony" as "all of Her Majesty's possessions abroad in which there shall exist a legislature." It then provided:

2. Any colonial law which is or shall be in any respect repugnant to the provisions of any Act of Parliament extending to the colony to which such law may relate, or repugnant to any order or regulation made under authority of such Act of Parliament, or having in the colony the force and effect of such Act, shall be read subject to such Act, order, or regulation, and shall, to the extent of such repugnancy, but not otherwise, be and remain absolutely void and inoperative.

3. No colonial law shall be or be deemed to have been void or inoperative on the ground of repugnancy to the law of England, unless the same shall be repugnant to the

provisions of some such Act of Parliament, order, or regulation as aforesaid.

37. Mr Crow submits that BIOT is a colony with a legislature, namely, the Commissioner. The Constitution Order is a law made for the Colony by Order in Council and therefore a "colonial law." It therefore cannot be void or inoperative by reason of its repugnancy to English common law doctrines of judicial review.

38. The Court of Appeal rejected this argument on the ground that the 1865 Act was concerned with the repugnancy of otherwise valid colonial laws to the law of England. The principles of judicial review, on the other hand, determined whether the Order in Council was valid in the first place. No question of repugnancy arose because, if the Order in Council was beyond the powers of Her Majesty in Council, there was no colonial law which could be repugnant to anything.

39. In a paper written for the Oxford Law Faculty (*Common Law Constraints: Whose Common Good Counts?* (<http://papers.ssrn.com/sol3/papers.cfm?abstract_id=1100628>) Professor Finnis of University College has persuasively argued that this is a slippery argument because repugnancy to English law (or fundamental principles of English law) can be regarded, and was regarded in the first half of the nineteenth century, as limiting the *powers* of colonial legislatures rather than as being an independent ground for invalidating laws otherwise validly made. I agree that a distinction between initial invalidity for lack of compliance with doctrines of English public law and invalidity for repugnancy to English law is too fine to be serviceable.

40. Nevertheless, I would reject the argument based on the Colonial Laws Validity Act for a different reason. In my opinion the Act was intended to deal with the validity of colonial laws (whether made by the local legislature or by Her Majesty in Council) from the perspective of their forming part of the local system of laws administered by the local courts. Section 3 made it clear that in considering the validity of such laws, the courts were not to concern themselves with the law of England, although they might well apply local principles of judicial review identical with those existing in English law. But these proceedings are concerned with the validity of the Order, not simply as part of the local law of BIOT but, as Professor Finnis says, as imperial

legislation made by Her Majesty in Council in the interests of the undivided realm of the United Kingdom and its non-self-governing territories. The Constitution Order created the BIOT legislature, in the form of the Commissioner, and it seems to me to illustrate the amphibious nature of the Order in Council, as both British and colonial legislation, that the legislature which is said to bring BIOT within the definition of a colony for the purposes of the Act was created by the very Order which is said to be a law "made for a colony." The fact is that Parliament in 1865 would simply not have contemplated the possibility of an Order in Council legislating for a colony as open to challenge in an English court on principles of judicial review. It was concerned with the law applicable by colonial courts, not English courts.

41. It therefore seems to me that from the point of view of the jurisdiction of the courts of the United Kingdom to review the exercise of prerogative powers by Her Majesty in Council, the Constitution Order is not a colonial law, although it may well have been from the point of view of a BIOT court applying BIOT law.

42. Sir Sydney's proposition that the Crown does not have power to remove an islander's right of abode in the Territory is in my opinion also too extreme. He advanced two reasons. The first was that a right of abode was a fundamental constitutional right. He cited the 29th chapter of Magna Carta:

> No freeman shall be taken, or imprisoned... or exiled, or any otherwise destroyed... but by the lawful judgment of his peers, or by the law of the land.

43. "But... by the law of the land" are in this context the significant words. Likewise Blackstone (*Commentaries on the Laws of England* (15th ed 1809) vol 1, p 137):

> But no power on earth, except the authority of Parliament, can send any subject of England out of the land against his will; no, not even a criminal.

44. That remains the law of England today. The Crown has no authority to transport anyone beyond the seas except by statutory authority. At common law, any subject of the Crown has the right to enter and remain in the United Kingdom whenever and for as long as he pleases: see *R v Bhagwan* [1972] AC 60. The Crown cannot remove this right by an exercise of the prerogative. That is because since the 17th century the prerogative has

not empowered the Crown to change English common or statute law. In a ceded colony, however, the Crown has plenary legislative authority. It can make or unmake the law of the land.

45. What these citations show is that the right of abode is a creature of the law. The law gives it and the law may take it away. In this context I do not think that it assists the argument to call it a constitutional right. The constitution of BIOT denies the existence of such a right. I quite accept that the right of abode, the right not to be expelled from one's country or even one's home, is an important right. General or ambiguous words in legislation will not readily be construed as intended to remove such a right: see *R v Secretary of State for the Home Department, Ex p Simms* [2000] 2 AC 115, 131–132. But no such question arises in this case. The language of section 9 of the Constitution Order could hardly be clearer. The importance of the right to the individual is also something which must be taken into account by the Crown in exercising its legislative powers—a point to which I shall in due course return. But there seems to me no basis for saying that the right of abode is in its nature so fundamental that the legislative powers of the Crown simply cannot touch it.

46. Next, Sir Sydney submitted that the powers of the Crown were limited to legislation for the "peace, order and good government" of the Territory. Applying the reasoning of the Divisional Court in *Bancoult (1)*, he said that meant that the law had to be for the benefit of the inhabitants, which could not possibly be said of a law which excluded them from the Territory.

47. There are two answers to this submission. The first is the prerogative power of the Crown to legislate for a ceded colony has never been limited by the requirement that the legislation should be for the peace, order and good government or otherwise for the benefit of the inhabitants of that colony. That is the traditional formula by which legislative powers are conferred upon the legislature of a colony or a former colony upon the attainment of independence. But Her Majesty exercises her powers of prerogative legislation for a non-self-governing colony on the advice of her ministers in the United Kingdom and will act in the interests of her undivided realm, including both the United Kingdom and the colony: see Halsbury's *Laws of England* (4th ed 2003 reissue) vol 6, para 716:

> The United Kingdom and its dependent territories within Her Majesty's dominions form one realm having one

undivided Crown... To the extent that a dependency has responsible government, the Crown's representative in the dependency acts on the advice of local ministers responsible to the local legislature, but in respect of any dependency of the United Kingdom (that is, of any British overseas territory) acts of Her Majesty herself are performed only on the advice of the United Kingdom government.

48. Having read Professor Finnis's paper, I am inclined to think that the reason which I gave for dismissing the cross-appeal in *R (Quark Fishing Ltd) v Secretary of State for Foreign and Commonwealth Affairs* [2006] 1 AC 529, 551 was rather better than the reason I gave for allowing the Crown's appeal and that on this latter point Lord Nicholls of Birkenhead was right.

49. Her Majesty in Council is therefore entitled to legislate for a colony in the interests of the United Kingdom. No doubt she is also required to take into account the interests of the colony (in the absence of any previous case of judicial review of prerogative colonial legislation, there is of course no authority on the point) but there seems to me no doubt that in the event of a conflict of interest, she is entitled, on the advice of Her United Kingdom ministers, to prefer the interests of the United Kingdom. I would therefore entirely reject the reasoning of the Divisional Court which held the Constitution Order invalid because it was not in the interests of the Chagossians.

50. My second reason for rejecting Sir Sydney's argument is that the words "peace order and good government" have never been construed as words limiting the power of a legislature. Subject to the principle of territoriality implied in the words "of the Territory", they have always been treated as apt to confer plenary law-making authority. For this proposition there is ample authority in the Privy Council (*R v Burah* (1878) 3 App Cas 889; *Riel v The Queen* (1885) 10 App Cas 675; *Ibralebbe v The Queen* [1964] AC 900) and the High Court of Australia *Union Steamship Company of Australia Pty Ltd v King* (1988) 166 CLR 1). The courts will not inquire into whether legislation within the territorial scope of the power was in fact for the "peace, order and good government" or otherwise for the benefit of the inhabitants of the Territory. So far as *Bancoult (1)* departs from this principle, I think that it was wrongly decided.

51. Sir Sydney placed great reliance upon a statement of Evatt J in *Trustees Executors and Agency Co Ltd v Federal Commissioner*

of *Taxation* (1933) 49 CLR 220, 234 that the question was "whether the law in question can be truly described as being for the peace, order and good government of the Dominion concerned." But this statement must not be wrenched from the context in which it was made. The judge was concerned with the principle of territoriality (the case was about whether Australian estate duty could be levied on movables situated abroad) and the emphasis was on the words "of the Dominion concerned." There was no suggestion that if the law satisfied the principle of territoriality (as this law and the Immigration Ordinance 1971 in *Bancoult (1)* obviously did) the courts could inquire into whether its objects could be said to be peace, order and good government.

52. Having rejected the extreme arguments on both sides, I come to what seems to me the main point in this appeal, namely the application of ordinary principles of judicial review. On this question there was a radical difference in the approaches advocated by the parties. Mr Crow said that because the Crown was acting in the interests of the defence of the realm, diplomatic relations with the United States and the use of public funds in supporting any settlement on the islands, the courts should be very reluctant to interfere. Judicial review should be undertaken with a light touch and the Order set aside only if it appeared to be wholly irrational. Sir Sydney, on the other hand, said that because the Order deprived the Chagossians of the important human right to return to their homeland, the Order should be subjected to a much more exacting test. As he said in his printed case (at para 137):

> Where a measure affects fundamental rights, or has profoundly intrusive effects, the courts will employ an 'anxious' degree of scrutiny in requiring the public body in question to demonstrate that the most compelling of justifications existed for such measures.

53. I would not disagree with this proposition, which is supported by a quotation from the judgment of Sir Thomas Bingham MR in *R v Ministry of Defence, Ex p Smith* [1996] QB 517, 554. However, I think it is very important that in deciding whether a measure affects fundamental rights or has "profoundly intrusive effects", one should consider what those rights and effects actually are. If we were in 1968 and concerned with a proposal to remove the Chagossians from their islands with little or no

provision for their future, that would indeed be a profoundly intrusive measure affecting their fundamental rights. But that was many years ago, the deed has been done, the wrong confessed, compensation agreed and paid. The way of life the Chagossians led has been irreparably destroyed. The practicalities of today are that they would be unable to exercise any right to live in the outer islands without financial support which the British Government is unwilling to provide and which does not appear to be forthcoming from any other source. During the four years that the Immigration Ordinance 2000 was in force, nothing happened. No one went to live on the islands. Thus their right of abode is, as I said earlier, purely symbolic. If it is exercised by setting up some camp on the islands, that will be a symbol, a gesture, aimed at putting pressure on the government. The whole of this litigation is, as I said in *R v Jones (Margaret)* [2007] 1 AC 136, 177 "the continuation of protest by other means." No one denies the importance of the right to protest, but when one considers the rights in issue in this case, which have to be weighed in the balance against the defence and diplomatic interests of the state, it should be seen for what it is, as a right to protest in a particular way and not as a right to the security of one's home or to live in one's homeland. It is of course true that a person does not lose a right because it becomes difficult to exercise or because he will gain no real advantage by doing so. But when a legislative body is considering a change in the law which will deprive him of that right, it cannot be irrational or unfair to consider the practical consequences of doing so. Indeed, it would be irrational not to.

54. My Lords, I think that if one keeps firmly in mind the practical effect of section 9 of the Constitution Order, the issues in this appeal fall into place. The government does not consider that it is in the public interest that an unauthorised settlement on the islands should be used as a means of exerting pressure to compel it to fund a resettlement which it has decided would be uneconomic. That is a view it is entitled to take. In the Court of Appeal, Sedley LJ treated the question of funding as irrelevant. The applicant was not asking for an order that the government fund resettlement. To focus on the logistics of resettlement was, he said, to miss the point:

> The point is that the two Orders in Council negate one of the most fundamental liberties known to human beings,

the freedom to return to one's own homeland, however poor and barren the conditions of life...

55. I respectfully think that this misses the point. Funding is the subtext of what this case is about. The Chagossians have, not unreasonably, shown no inclination to return to live Crusoe-like in poor and barren conditions of life. The action is, like *Bancoult (1)*, a step in a campaign to achieve a funded resettlement. The attempt to achieve that through domestic litigation foundered before Ouseley J. But that does not mean that the Secretary of State is bound to assume that these expensive proceedings are purely academic. The Secretary of State is surely entitled to take into account that once a vanguard of Chagossians establishes itself on the islands in poor and barren conditions of life, there may be a claim that the United Kingdom is subject to a sacred trust under article 73 of the United Nations Charter to "ensure... [the] economic, social and educational advancement" of the residents and to send reports to the Secretary-General.

56. It is true that the Chagossians will now require immigration consent even to visit the islands. But the government have made it clear that such visits, to tend graves and so forth, will be allowed, and since in practice they are funded by the BIOT administration, immigration consent will be no more than an additional formality. Furthermore, there is no reason why, if at some time in the future, circumstances should change, the controls should not be lifted.

57. In addition, as Mr Rammell told the House of Commons, the government had to give due weight to security interests. The United States had expressed concern that any settlement on the outer islands would compromise the security of its base on Diego Garcia. A representative of the State Department wrote a letter for use in these proceedings, giving details of the ways in which it was feared that the islands might be useful to terrorists. Some of these scenarios might be regarded as fanciful speculations but, in the current state of uncertainty, the government is entitled to take the concerns of its ally into account.

58. Policy as to the expenditure of public resources and the security and diplomatic interests of the Crown are peculiarly within the competence of the executive and it seems to me quite impossible to say, taking fully into account the practical interests of the Chagossians, that the decision to reimpose immigration control on the islands was unreasonable or an abuse of power.

59. The applicant's alternative ground for judicial review was that the Foreign Secretary's press announcement after the judgment in *Bancoult (1)*, accompanied by the revocation of immigration controls by the 2000 Ordinance, was a promise which created a legitimate expectation that the islanders would be free from such controls. In the absence of a change in relevant circumstances, the Crown should be required to keep its promise.

60. The relevant principles of administrative law were not in dispute between the parties and I do not think that this is an occasion on which to re-examine the jurisprudence. It is clear that in a case such as the present, a claim to a legitimate expectation can be based only upon a promise which is "clear, unambiguous and devoid of relevant qualification": see Bingham LJ in *R v Inland Revenue Commissioners, Ex p MFK Underwriting Agents Ltd* [1990] 1 WLR 1545, 1569. It is not essential that the applicant should have relied upon the promise to his detriment, although this is a relevant consideration in deciding whether the adoption of a policy in conflict with the promise would be an abuse of power and such a change of policy may be justified in the public interest, particularly in the area of what Laws LJ called "the macro-political field": see *R v Secretary of State for Education and Employment, Ex p Begbie* [2000] 1 WLR 1115, 1131.

61. In my opinion this claim falls at the first hurdle, that is, the requirement of a clear and unambiguous promise. The Foreign Secretary said that the Crown accepted the decision in *Bancoult (1)* that the 1971 Immigration Ordinance was outwith the powers the BIOT Order and that a new Ordinance would be made which would allow "the Îlois to return to the outer islands." This was done. Nothing was said about how long that would continue. But the background to the statement was the ongoing study "on the feasibility of resettling the Îlois." If that resulted in a decision to resettle, then one would expect the right of abode of the Chagossians on the outer islands to continue. On the other hand, if it did not, the whole situation might need to be reconsidered. It was obvious that no one contemplated the resettlement of the Chagossians unless the government, taking into account the findings of the feasibility study, decided to support it. If they did not, a new situation would arise. The government might decide that little harm would be done by leaving the Chagossians with a theoretical right to return to the islands and for two years after the feasibility report, that seems to have

been the view that was taken. But the Foreign Secretary's press statement contained no promises about what, in such a case, would happen in the long term.

62. No doubt the Chagossians saw things differently. As we have seen, they tried to persuade the government that the press statement amounted to the adoption of a policy of resettlement. They realised that what mattered was whether the government was willing to fund resettlement. Otherwise they had secured an empty victory. But the question is what the statement unambiguously promised and in my opinion it comes nowhere near a promise that, even if there could be no resettlement, immigration control would not be reimposed.

63. Even if it could be so construed, I consider that there was a sufficient public interest justification for the adoption of a new policy in 2004. For this purpose it is relevant that no one acted to their detriment on the strength of the statement, that the rights withdrawn were not of practical value to the Chagossians and that the decision was very much concerned with the "macro-political field."

64. That leaves two points which were not considered by the Divisional Court or the Court of Appeal and which were lightly touched upon in argument but upon which the House is invited to rule. They are whether, in principle, the validity of the Constitution Order may be affected by the Human Rights Act 1998 or by international law. I do not think that the Human Rights Act 1998 has any application to BIOT. In 1953 the United Kingdom made a declaration under article 56 of the European Convention on Human Rights extending the application of the Convention to Mauritius as one of the "territories for whose international relations it is responsible." That declaration lapsed when Mauritius became independent. No such declaration has ever been made in respect of BIOT. It is true that the territory of BIOT was, until the creation of the colony in 1965, part of Mauritius. But a declaration, as appears from the words "for whose international relations it is responsible" applies to a political entity and not to the land which is from time to time comprised in its territory. BIOT has since 1965 been a new political entity to which the Convention has never been extended.

65. If the Convention has no application in BIOT, then the actions of the Crown in BIOT cannot infringe the provisions of the

Human Rights Act 1998: see *R (Quark Fishing Ltd) v Secretary of State for Foreign and Commonwealth Affairs* [2006] 1 AC 529. The applicant points out that section 3 of the BIOT Courts Ordinance 1983 provides that the law of England as in force from time to time shall apply to the territory. So, they say, the Human Rights Act, when enacted, became part of the law of the territory. So be it. But the Act defines Convention rights (in section 21(1)) as rights under the Convention "as it has effect for the time being in relation to the United Kingdom." BIOT is not part of the United Kingdom and the Human Rights Act, though it may be part of the law of England, has no more relevance in BIOT than a local government statute for Birmingham.

66. As for international law, I do not understand how, consistently with the well-established doctrine that it does not form part of domestic law, it can support any argument for the invalidity of a purely domestic law such as the Constitution Order.

67. I would allow the appeal, set aside the orders of the Divisional Court and the Court of Appeal and dismiss the application.

NOTES

Preface

1. M. Ducasse, "Île va sang dire...", *Mélangés* I (Beau-Bassin/Mauritius: Éditions Vilaz Métiss 2002).
2. N. Chomsky, *Hegemony or Survival: America's Quest for Global Dominance* (London: Penguin 2004) p. 162.
3. A.R. Lewis, *The American Culture of War: The History of U.S. Military Force from World War II to Operation Iraqi Freedom* (London: Routledge 2007) p. 450.
4. P. Saddul (ed.), *Philip's Atlas of Mauritius* (Port Louis: Éditions de l'Océan Indien 2007) p. 5.

1 History: Empire's Last-Born Colony

1. Diego García (1471–1535); *Enciclopédia Universal Ilustrada* vol. 25 (Barcelona: Espasa 1924) p. 758, and *Grande Enciclopédia Portuguesa e Brasileira* vol. 12 (Lisbon 1940) p. 162. In Gerhard Mercator's famous atlas of the world, *Nova et Aucta Orbis Terrae Descriptio ad Usum Navigantium Emendate Accommodata* (Duisburg 1569), the island was shown as *Isola de Don García*.
2. Following the capitulation of Isle de France (December 3, 1810) and the Peace Treaty of Paris (May 30, 1814) which ended the Napoleonic Wars, Britain took possession of Mauritius and the adjoining islands; see C. Parry (ed.), *Consolidated Treaty Series* vol. 63 (1969) p. 171, and R. Scott, *Limuria: The Lesser Dependencies of Mauritius* (London: Oxford University Press 1961) p. 105.
3. See E.P. Hoyt, *The Last Cruise of the Emden* (New York: Macmillan 1966) pp. 117–119; P.G. Huff, *S.M.S. Emden* (Kassel: Hamecher 1994) pp. 162, 194; and R. Edis, *Peak of Limuria: The Story of Diego Garcia and the Chagos Archipelago* (London: Chagos Conservation Trust, rev. edn. 2004) pp. 53–56.
4. See "Chagos Archipelago: Coconuts and Turtles," *Times* (London, May 25, 1920) p. 35. The Paris-based *Société Huilière de Diego et Peros* (Diego Ltd.) since 1941), which had originally operated the plantation and three copra oil factories on the island, was sold to the Chagos-Agalega Co. of the Seychelles in 1962; A. Toussaint, *Histoire des Îles Mascareignes* (Paris: Berger-Levrault 1972) p. 273/footnote 1.
5. See the mission reports by R. Dussercle, *Archipel de Chagos* (Port Louis: General Printing 1934–1937).
6. As recorded in the *Reminiscences of Admiral Roy L. Johnson* (Annapolis/MD: U.S. Naval Institute Oral History Transcript 1982) p. 336. On the role of Barber, see also M. Bezboruah,

U.S. Strategy in the Indian Ocean: The International Response (New York: Praeger 1977) pp. 53, 227; V.B. Bandjunis, *Diego Garcia: Creation of the Indian Ocean Base* (San José/CA: Writer's Showcase 2001) pp. 2, 22; and David Vine, Empire's Footprint: Expulsion and the U.S. Military Base on Diego Garcia (PhD Dissertation: City University of New York 2006) pp. 9, 30, 134–141; revised as *Island of Shame: The Secret History of the U.S. Military Base on Diego Garcia* (Princeton/NJ: Princeton University Press 2009).

7. J. Pilger, *Freedom Next Time: Resisting the Empire* (New York: Nation Books 2007) pp. 20–61, p. 23.
8. Bandjunis (note 6 above) pp. 11–13 and 24.
9. Aldabra, an uninhabited atoll in the Seychelles (World Heritage site since 1982), is home to a unique population of giant tortoises and other endangered species. See D.R. Stoddart (ed.), "Ecology of Aldabra Atoll, Indian Ocean," *Atoll Research Bulletin* vol. 118 (1967) pp. 1–141; *id.*, "Scientific Studies at Aldabra and Neighbouring Islands," *Philosophical Transactions of the Royal Society of London, Series B: Biological Sciences* vol. 260 No. 836 (1971) pp. 5–29; and T. Dalyell, "Holding Policy-Makers to Account: The Problem of Expertise," in C. Hill and P. Beshoff (eds.), *Two Worlds of International Relations: Academics, Practitioners and the Trade in Ideas* (London: Routledge 1994) pp. 118–135, at p. 119. It was British MP Tom Dalyell who had led the campaign against militarization of Aldabra Island in the House of Commons; see T. Beamish, *Aldabra Alone* (London: Allen & Unwin 1970) pp. 180–206; and C.T. Sandars, *America's Overseas Garrisons: The Leasehold Empire* (Oxford: Oxford University Press 2000) p. 56.
10. As quoted by Bandjunis (note 6) p. 14, who also relates (at p. 2) that it was Admiral Rivero who had been first to propose, in May 1960, to detach the Chagos Archipelago from British Mauritius and to secure rights of use for the United States. See also Vine (note 6 above) pp. 10, 139; and the *Reminiscences of Admiral Horacio Rivero Jr.* (Annapolis/MD: U.S. Naval Institute Oral History Transcript vol. 3, May 1978).
11. Admiral John S. McCain (1884–1945), commander of a U.S. carrier task force in the Pacific during World War II, as quoted in A. Bhatt, *The Strategic Role of Indian Ocean in World Politics: The Case of Diego Garcia* (New Delhi: Ajanta 1992) p. 7 and in J.R. Mancham, *War on America Seen from the Indian Ocean* (St. Paul/MN: Paragon 2001) p. 44. See generally A.S. Erickson, W.C. Ladwig III, and J.D. Mikolay, *Diego Garcia's Strategic Past, Present and Future*, paper presented at the annual meeting of the American Political Science Association (Boston/MA, August 28, 2008) and *id.*, "The Military Geography of Diego Garcia: Future Implications for U.S. Power Projection in the Indian Ocean," in C. Lord and A. Erickson (eds.), *U.S. Basing and Presence in the Asia-Pacific* (Newport/RI: Naval War College Press, forthcoming 2010), which also includes an analysis of the current position of India and China with regard to Diego Garcia.
12. Inscription prominently stenciled on the highest water tower near the harbor entrance of the U.S. naval base at Diego Garcia: "Welcome to the Footprint of Freedom"; see S. Winchester, *Outposts: Journeys to the Surviving Relics of the British Empire* (London: Penguin Books 2003) p. 47; Edis (note 3) pp. 3 and 86; and the video produced by the U.S. Navy, <www.youtube.com/watch?v=uWYCcdiqsBA> (accessed 22 July 2008). See generally D.R. Stoddart and J. Taylor (eds.), "Geology and Ecology of Diego Garcia Atoll," *Atoll Research Bulletin* No. 149 (1971) pp. 1–149; I.B. Walker, *The Complete Guide to the Southwest Indian Ocean* (Argelès-sur-Mer: Cornelius Books 1993) pp. 561–572; and the references in Chapter 5 (note 2).
13. See A. Oraison, "À propos du litige anglo-mauricien sur l'archipel des Chagos (la succession d'États sur les îles Diego Garcia, Peros Banhos et Salomon)," *Revue Belge de Droit International* vol. 23 (1990) pp. 5–53.
14. "Special Committee on the Situation with Regard to Implementation of the Declaration on the Granting of Independence to Colonial Countries and Peoples," established by the UN General Assembly in November 1961 to succeed the "Special Committee on Information from Non-Self-Governing Territories," established by UNGA Resolution 146 (II) on

November 3, 1947, and renamed by UNGA Resolution 569(V) of January 18, 1952. See pp. 5, 17, 24, and U.O. Umozurike, *Self-Determination in International Law* (Hamden/CT: Archon 'Books 1972) at p. 74 and text at note 30; and in Chapter 2 (notes 6 and 21).

15. UNGA Resolution 1514 (XV) of December 14, 1960, UN Document A/38/711. On the political embarrassment of U.S. abstention see D.A. Kay, "The United Nations and Decolonization," in J. Barros (ed.), *The United Nations: Past, Present and Future* (New York: Free Press 1972) pp. 143–170, at p. 152.

16. As quoted by Lord Justice [Sir John Grant McKenzie] Laws, in *The Queen (ex parte Bancoult) vs. Foreign and Commonwealth Office*, *Law Reports: Queen's Bench Division* [2001] p. 1080; see also Chapter 2 (note 47) and S. Winchester, "Diego Garcia," *Granta: The Magazine of New Writing* No. 73 (2001) pp. 207–226, at p. 211. The secret UK-U.S. negotiations on Diego Garcia since 1963 are simply documented in N.D. Howland (ed.), 'Indian Ocean', *Foreign Relations of the United States, 1964–1968*, vol. 21 (Washington, D.C.: U.S. Government Printing Office 2000) pp. 83–117.

17. Pilger (note 7) p. 24.

18. Oraison (note 13) p. 57; and *id.*, "Le contentieux territorial anglo-mauricien sur l'archipel des Chagos revisité," *Revue de Droit International et de Sciences Diplomatiques et Politiques* vol. 83 (2005) pp. 109–208, at p. 151. [Pound-dollar conversion rates are as of the referenced time.] Initially, the Mauritian negotiators had also asked for UK and U.S. commitments to buy all their sugar at a preferential price and to accept Mauritian immigrants, but those demands were eventually dropped; see Bandjunis (note 6) p. 15.

19. M. Denuzière, "Les Seychelles: au plus près du bonheur—demain l'indépendance," *Le Monde* (Paris, 26 May 1976) p. 6; and Mancham (note 11) p. 42. The land on *Aldabra* island was crown property already; on the *Chagos* land deal, see Chapter 2 (note 10).

20. See J.C. de l'Estrac (Chairman), *Report of the Select Committee on the Excision of the Chagos Archipelago*, Mauritius Legislative Committee (Port Louis: Government Printer 1983) p. 4; A.S. Simmons, *Modern Mauritius: The Politics of Decolonization* (Bloomington: University of Indiana Press 1982) p. 173; J. Houbert, "The Indian Ocean Creole Islands: Geo-Politics and Decolonization," *Journal of Modern African Studies* vol. 30 (1992) pp. 465–484, at pp. 466–475; Oraison (note 13) pp. 27–28. The Mauritian chief delegate, Prime Minister Seewoosagur Ramgoolam, was knighted by the queen in 1966.

21. That commitment has repeatedly been confirmed by the United Kingdom; see note 40.

22. British Indian Ocean Territory Order 1965, Statutory Instruments [1965] No. 1920, as amended in Statutory Instruments [1968] No. 111; *Halsbury's Laws of England*, 4th edn. (London: LexisNexis 2003 reissue) vol. VI, para. 854. See C. Rousseau, "Chronique des faits internationaux: Grande-Bretagne—création d'une nouvelle colonie britannique dans l'Océan Indien par l'ordre en conseil du 7 novembre 1965," *Revue Générale de Droit International Public* vol. 70 (1966) p. 171; R. Aldrich and J. Connell, *The Last Colonies* (Cambridge: Cambridge University Press 1998) pp. 178–182.

23. L. Moor and A.W.B. Simpson, "Ghosts of Colonialism in the European Convention on Human Rights," *British Yearbook of International Law* vol. 26 (2005) pp. 121–193, at p. 162 even suggest that BIOT "was invented with a view to violating the rights" of the indigenous inhabitants of the Chagos islands. See also Chapter 2 (notes 69–77).

24. De l'Estrac Report (note 20) pp. 14, 36; see also S.S. Harrison, "India, the United States, and Superpower Rivalry in the Indian Ocean," in S.S. Harrison and K. Subrahmanyam (eds.), *Superpower Rivalry in the Indian Ocean: Indian and American Perspectives* (New York: Oxford University Press 1989) pp. 246–286, at p. 257.

25. See F. Wooldridge, "*Uti Possidetis* Doctrine," in R. Bernhardt (ed.), *Encyclopedia of Public International Law* vol. 4 (Amsterdam: Elsevier 2000) pp. 1259–1262, and S. Lalonde, "*Uti Possidetis*: Its Colonial Past Revisited," *Revue Belge de Droit International* vol. 34 (2001) pp. 23–90.

26. UNGA Resolution 2066 (XX) of December 16, 1965 (*Question of Mauritius*), UN Document A/6014/57.

27. 603 United Nations Treaty Series 273 (No. 8737), 18 United States Treaties 28 (TIAS No. 6196), United Kingdom Treaty Series [1967] No. 15 (Cmnd. 3231), reprinted in K.S. Jawatkar, *Diego Garcia in International Diplomacy* (Bombay: Popular Prakash 1983) p. 303, in Bandjunis (note 6) p. 263, and in appendix I; (p. **69**) as amended in 1976 (see note 31). See also J. Woodliffe, *The Peacetime Use of Foreign Military Installations under Modern International Law* (Dordrecht: Nijhoff 1992) pp. 89–90, and A. Cooley, *Base Politics: Democratic Change and the U.S. Military Overseas* (Ithaca/NY: Cornell University Press 2008) p. 41. In reality, the agreement and its secret side-notes (concluded simultaneously with a UK-U.S. Agreement on Space Tracking Facilities in the Seychelles) were probably *not* signed on December 30, 1966, but at least one day earlier, as shown on the copies filed with the U.S. National Archives (note 37), which were certified by U.S. consul R.E. White Jr. in London on December 29, 1966. The tactical purpose of this curious postdating apparently was to delay distribution of the documents to Parliament and Congress until after the New Year's recess, so as to minimize unwanted political attention. According to C. Rousseau, "Chronique des faits internationaux," *Revue Générale de Droit International Public* vol. 71 (1967) p. 1101, the text of the agreement was not made public until April 27, 1967; it was registered with the UN Treaty Section in New York (by the UK Foreign Office) on August 22, 1967.
28. Concluded at London (October 24, 1972), 866 United Nations Treaty Series 302, 23 United States Treaties 3087 (TIAS No. 7481), United Kingdom Treaty Series [1972] No. 126 (Cmnd. 5160), reprinted in Bandjunis (note 6) p. 277 and appendix III (p. **84**); London (February 25, 1976), 1018 United Nations Treaty Series 372, 27 United States Treaties 315 (TIAS No. 8230), United Kingdom Treaty Series [1976] No. 19 (Cmnd. 6413), Bandjunis p. 287, and appendix IV (p. **91**); Washington, D.C. (December 13, 1982), 2001 United Nations Treaty Series 397, 24 United States Treaties 4553 (TIAS No. 10616), appendix VI (p. **107**); Washington, D.C. (November 16,1987), 1576 United Nations Treaty Series 179 (No. 27519), United Kingdom Treaty Series [1988] No. 60, appendix VII (p. **109**); London (July 21, 1999), 2106 United Nations Treaty Series 294 (No. 36627), United Kingdom Treaty Series [2000] No. 1 (Cmnd. 4582), appendix VIII (p. **111**). The subsequent exchanges of letters (of September 10, 2001; July 29, 2002; and December 7, 2004, respectively; appendix IX, p. **116**) were unpublished, and texts were made available to the author by the UK Foreign and Commonwealth Office under freedom-of-information request 0670–08 (September 2, 2008); see also the written statement by the Foreign Secretary in *House of Commons Debates*, vol. 424 (July 21, 2004), col. 340W, *British Yearbook of International Law* vol. 75 (2004) p. 672.
29. See Chapter 5 (note 83).
30. Note 14; see A. Rigo Sureda, *The Evolution of the Right of Self-Determination: A Case Study of U.N. Practice* (Leiden: Sijthoff 1973) p. 202 footnote 1. For example, during its session in New York in July–August 1970, the Committee of Twenty-Four "confirmed that the detachment of a number of islands from the Seychelles by the administrative Power, and the setting up of the so-called British Indian Ocean Territory with the purpose of establishing a military base in that Territory jointly with the United States, was incompatible with the United Nations Charter and the Declaration on Decolonization" and "again called on the administrative Power to respect the territorial integrity of the Seychelles and to return immediately to that Territory the islands detached from it in 1965"; *UN Monthly Chronicle* vol. 7 No. 8 (1970) pp. 54–55. See also P.M. Allen, *Self-Determination in the Western Indian Ocean*, International Conciliation No. 560 (Washington, D.C.: Carnegie Endowment 1966) p. 4.
31. Statutory Instruments [1976] No. 893; Diego Garcia Amended Agreement of June 25, 1976 (effective June 28, 1976), 1032 United Nations Treaty Series 323, United Kingdom Treaty Series [1976] No. 88 (Cmnd. 6610), 27 United States Treaties 3448 (TIAS No. 8376), Bandjunis (note 6) p. 305, and appendix V (p. **105**). The current BIOT commissioner is Colin Roberts (appointed in June 2008, also commissioner for the British Antarctic Territory), assisted by an administrator (Joanne Yeadon, appointed in December 2007, also administrator for the Pitcairn Islands). The current resident representative of the

commissioner (and justice of the peace) in Diego Garcia (appointed in October 2007) is Cdr. Gary L. Brooks; the current U.S. commanding officer in Diego Garcia (appointed in June 2008) is Capt. Daniel T. McNamara.

32. Hearings before the Special Subcommittee on Investigations of the Committee on International Relations, House of Representatives, 94th Congress, 1st Session, June 5 and November 4, 1975 (Washington, D.C.: U.S. Government Printing Office 1975), pp. 63–320; see also Chapter 2 (note 26).

33. As of September 19, 1975, British "detachment costs" incurred were listed as totaling $31.33 million, including $8.4 million to Mauritius for loss of sovereignty, $3.75 million to the *Chagos-Agalega* copra plantations for transfer of freehold, $17.36 million to the Seychelles for construction of Mahé airport, and $1.82 million to Mauritius for relocation of inhabitants; see U.S. General Accounting Office (GAO, now Government Accountability Office), *Financial and Legal Aspects of the Agreement on the Availability of Certain Indian Ocean Islands for Defense Purposes*, Report of the Comptroller General of the United States, B-184915 (Washington, D.C.: GAO, January 7, 1976) p. 1. In response to parliamentary questions in 1975 and 2002, the UK Foreign Office stated that the United States had agreed to contribute up to £5 million ($14 million) "towards the costs of setting up the BIOT as a separate entity"; see FCO Minister of State David H. Ennals [later Baron Ennals of Norwich], *House of Commons Debates* vol. 898 col. 128-32W (October 21, 1975); and *British Yearbook of International Law* vol. 73 (2002) p. 700.

34. US-UK POLARIS Sales Agreement signed at Washington, D.C. on April 6, 1963, 474 United Nations Treaty Series 49, 14 United States Treaties 321 (TIAS No. 5313), United Kingdom Treaty Series [1963] No. 59. An (unpublished) "classified minute" to the agreement is referred to in para. 3(b) of the side-note to the 1966 UK-US Agreement on Diego Garcia (see p. **7**).

35. Personally authorized by U.S. Secretary of Defense Robert S. McNamara, memorandum of June 14, 1965, to Secretary of the Air Force Eugene Zuckert; reprinted in Howland (note 16) p. 97. See Bandjunis (note 6) pp. 28 and 132–133 (table 4), Sandars (note 9) p. 57; and R.B. Rais, *The Indian Ocean and the Superpowers: Economic, Political and Strategic Perspectives* (Towota/NJ: Barnes & Noble 1987) p. 78 (table 5.1).

36. See the GAO report (note 33) p. 4; and Bandjunis (note 6) p. 133. According to confidential telegram No. 1318 dated February 18, 1970, from the U.S. embassy in London to the State Department, the UK Foreign Office "having well and truly cooked its books vis-à-vis Parliament on BIOT financing" did not wish to make the deal public; so the State Department assured them that the matter would be discussed "in executive session only" [which was not true]; see Nixon Presidential Materials, NSC files, box 726, country files Europe/United Kingdom vol. II, as quoted in L.W. Qaimmagami & A.H. Howard (eds.), *Foreign Relations of the United States, 1969-1976*, vol. 24 (Washington, D.C.: U.S. Government Printing Office 2008) p. 126/footnote 2.

37. Letter (Note No. 26, marked "secret") dated December 30, 1966. from David K.E. Bruce CBE (US ambassador in London) to UK foreign secretary George Brown [later Baron George-Brown of Jevington, PC], confirmed by the UK Minister of State for Foreign Affairs [Alun Jones, Baron Chalfont of Llantaman, OBE MC PC] on the same date (Letter No. *AU 1199*, marked "secret"); with attached *Agreed Minute* dated December 30, 1966 (marked "confidential," appendix II, p. **81**). The letters/attachments were filed in the US National Archives and Records Administration, Washington/DC, file RG 59/150/64-65 (1964–1966, Box No. 1552) and declassified on November 16, 2005, but have since been removed by the US Government "for national security reasons." They are still on file with the UK National Archive, FO 93/8/401.

38. According to Article 111 of the Mauritian Constitution, as amended by Act No. 48 of December 17, 1991, "Mauritius includes…the Chagos Archipelago, including Diego Garcia"; *cf.* A. Chellapermal, *The Problem of Mauritius Sovereignty Over the Chagos Archipelago*

and the Militarization of the Indian Ocean (Perth: University of Western Australia Press 1984).
39. E.g., see the statements by Mauritius to the UN General Assembly on December 5, 1983, October 9, 1987, October 12, 1988, and November 11, 2001; Official Records, UN Documents A/38/711 p. 1, A/42/32 p. 48, A/43/28 p. 38, A/56/46 p. 17; *Revue Générale de Droit International Public* vols. 92 (1988) p. 701, 94 (1989) p. 493; and J.C. de l'Estrac, "Diego Garcia: Mauritius Battles a Superpower to Reclaim a Cold War Hostage," *Parliamentarian: Journal of the Parliaments of the Commonwealth* vol. 72:4 (1991) pp. 267–270. These declarations contradict wishful statements to the effect that the BIOT excision "appears to have been accepted, at least as a temporary measure"; J. Crawford, *The Creation of States in International Law* (2nd edn. Oxford: Clarendon 2006) p. 337. See also T. Ollivry, *Diego Garcia: enjeux stratégiques, diplomatiques et humanitaires* (Paris: Harmattan 2008) p. 76.
40. E.g., in the UN General Assembly in 1983 and 2002, see *British Yearbook of International Law* vol. 55 (1984) p. 519, and vol. 73 (2002) p. 701; and in the UK House of Commons in 1990, 1992, 1998, 2000, 2001, 2002, 2004, and 2006; see *British Yearbook of International Law* vol. 61 (1990) p. 572, vol. 69 (1998) p. 540, vol. 71 (2000) p. 592, vol. 72 (2001) p. 633, vol. 73 (2002) p. 707, vol. 75 (2004) p. 666, and vol. 77 (2006) p. 639. See also Chapter 6 (note 34).
41. After bilateral conversations in London (on the occasion of a Commonwealth Heads of Government Meeting on Reform of International Institutions, June 9–10, 2008), "the British Prime Minister took note of the concerns expressed and proposed that the issues would be considered, in the first instance, during forthcoming talks between the UK and Mauritius at the designated officials' level"; Cabinet Decisions: June 20, 2008, Mauritius Prime Minister's Office Homepage <http://www.gov.mu/portal/goc/pmo/file/cabinet-pmo.jsp>.
42. Statement by US Secretary of State Madeleine K. Albright during her visit to Mauritius on December 10, 2000, *Department of State Press Transcript*, p. 3.
43. Convention (and Protocols) for the Protection, Management and Development of the Marine and Coastal Environment of the Eastern Africa Region, adopted in Nairobi on June 21, 1985, under the auspices of the United Nations Environment Programme (UNEP), in force May 30, 1996; text in *Official Journal of the European Communities* [1986] C 253 p. 10, and in P.H. Sand, *Marine Environment Law in the United Nations Environment Programme* (London: Cassell Tycooly 1988) pp. 156–190. Mauritius has made it clear that its ratification (of July 3, 2000) also applies to the Chagos Archipelago; see p. **66**, and *Nairobi Convention: National Status Report on the Marine and Coastal Environment* (Port Louis: Ministry of Environment and National Development Unit, October 2007) pp. 14, 23. The Mauritian Government takes the position that territorial application of treaties to the Chagos is implied by virtue of Article 111 of the Mauritian Constitution (note 38); e.g., its ratification of the 1992 Rio Convention on Biological Diversity (1760 United Nations Treaty Series 79) on September 4, 1992 (see Chapter 5, note 8). In some cases, such as when acceding to the 1954 Hague Convention for the Protection of Cultural Property in the Event of Armed Conflict (249 United Nations Treaty Series 240), Mauritius explicitly declared (on September 22, 2006) that the convention shall also extend to "the Chagos Archipelago, including Diego Garcia."
44. Convention on the Transfer of Sentenced Persons, 1496 United Nations Treaty Series 91; see the UK protest of January 28, 2005, against the Mauritian declaration of June 18, 2004 (reiterated on 14 April 2005) which extended to the Chagos Archipelago, including Diego Garcia, its accession to that convention (which the UK had previously declared applicable to BIOT on January 21, 1987); Council of Europe Treaty Office, List of Declarations with respect to Treaty No. 112 (Strasbourg: Council of Europe 2008).
45. See the Mauritian protest against the UK ratifying on behalf of BIOT (on May 15, 1987) the 1985 Vienna Convention for the Protection of the Ozone Layer and the 1987 Montreal Protocol on Ozone-Depleting Substances, *British Yearbook of International Law*

vol. 64 (1993) p. 639, and the UK protest against Mauritius acceding to those same treaties on behalf of the Chagos Archipelago (on August 18, 1992), *British Yearbook of International Law* vol. 68 (1997) p. 485. The UK also extended to BIOT its ratification of the 1973 Convention on International Trade in Endangered Species of Wild Fauna and Flora (CITES, 993 United Nations Treaty Series 243) and of the 1979 Migratory Species Convention (Chapter 4, note 51); but see p. 51 on *non*-extension to BIOT of the UK ratification of the 1992 Biodiversity Convention. See also the Mauritian declaration of September 24, 2003, objecting to the UK declaration of May 2, 1979, which extended to BIOT its ratification of the 1973 Convention on the Prevention and Punishment of Crimes Against Internationally Protected Persons (1035 United Nations Treaty Series 167); and notes 49 and 61. On the extension to BIOT (on October 26, 2005) of the UK ratification of the 1992 Chemical Weapons Convention (already ratified by Mauritius on February 9, 1993), see note 73.

46. British Indian Ocean Territory, *Environment (Protection and Preservation) Zone* proclaimed on September 17, 2003, with geographical boundaries notified to the UN Secretariat (on March 12, 2004) which are identical to those of the BIOT *Fisheries Conservation and Management Zone* proclaimed by the Fishery Limits Ordinance of October 1, 1991; see *British Yearbook of International Law* vol. 62 (1991) p. 648, *cf.* vol. 74 (2003) p. 680; and *United Nations Law of the Sea Bulletin* No. 54 (2004) p. 99. According to a ministerial statement in the *House of Lords Debates* vol. 659 (March 31, 2004) pp. 240331–30, the designation was made "under Article 75 of UNCLOS," i.e., as an "exclusive economic zone"; see the map in Sheppard and Spalding (Chapter 5, note 59) p. v, and P.H. Sand, "Green Enclosure of Ocean Space: *Déjà Vu?*," *Marine Pollution Bulletin* vol. 54 (2007) pp. 374–376, at p. 375. On objections by the Maldives, see pp. **9, 32, 67** and note 50.

47. See the Agreement between the European Economic Community and the Government of Mauritius on Fishing in Mauritian Waters, of June 10, 1989, Official Journal of the European Communities [1989] L 159/2, which in its preamble recalls the Mauritian declaration of a 200-mile EEZ in accordance with the Law of the Sea Convention, and in Art. 1 defines the waters of Mauritius as "the waters over which Mauritius has sovereignty or jurisdiction in respect of fisheries... in accordance with the provisions of the United Nations Convention on the Law of the Sea"; Art. 10 confirms that the Agreement applies "to the territory of Mauritius" (see note 38). A UK/Mauritius declaration on the conservation of fisheries signed on January 27, 1994 (reiterated on March 14, 1997) simply restates the countries' reciprocal (competing) claims of sovereignty over the Chagos islands "and the surrounding maritime areas"; see *British Yearbook of International Law* vol. 65 (1994) p. 582, vol. 68 (1997) p. 477. However, only the BIOT Administration collects fishery licence fees for the Chagos EEZ from foreign commercial fishing companies (Spanish, French, and Japanese)—totalling over £4 million (approx. $6 million) since 2003; see written answer, *House of Commons Debates* vol. 446 (May 22, 2006) col. 1415W.

48. Southern Indian Ocean Fisheries Agreement (SIOFA), adopted under the auspices of the Food and Agriculture Organization of the United Nations (FAO) at Rome on July 7, 2006, not yet in force; text in Official Journal of the European Union [2006] L196/51. In a declaration made upon signature, "Mauritius reiterates its rights to exercise complete and full sovereignty over its territory, including the territory and maritime zones of the Chagos Archipelago and Tromelin as defined in the Constitution of Mauritius." Article 3 of the Agreement defines the geographical area of application as including the Chagos Archipelago [within FAO Statistical Area No. 51]; however, a "fishing area map" annexed to Appendix 3 of the Final Act of the Rome Conference clearly excludes the EEZ around the archipelago—effectively "black-holing" it. The *de facto* deletion of BIOT from the SIOFA Agreement—presumably dictated by the political-military priorities of UK-US cooperation in Diego Garcia—is particularly unfortunate in view of the urgent need for

regional cooperation to cope with the threat of illegal fishing in the archipelago, especially by Sri Lankan shark-fin poachers; see M.D. Spalding, "Partial Recovery of Sharks in the Chagos Archipelago," *Shark News* vol. 15 (2004) pp. 12–13.
49. Ratification by Mauritius on November 4, 1994; accession by the United Kingdom on July 25, 1997, extended to BIOT. When acceding to the related 1995 Implementation Agreement on Straddling Fish Stocks, Mauritius objected (by a declaration on March 25, 1997) to Britain's extension of the agreement to BIOT, reaffirming "its sovereignty and sovereign rights over these islands, namely the Chagos Archipelago which form an integral part of the national territory of Mauritius, and over their surrounding maritime spaces." The UK defended its position by a declaration on July 30, 1997, and (following UK ratification of the agreement on behalf of BIOT, on December 3, 1999) Mauritius reiterated its objection by a further declaration on February 8, 2002.
50. Notification of November 29, 2007; see p. **32**, and M. O'Shea, "Serious Questions Over Sea Boundaries Between Maldives and British Indian Ocean Territory," <http://www.maldivesculture.com> (on-line December 20, 2007). For a location map showing the previous (equidistant) partition of the EEZ between the Maldives and Chagos, see M. Shiham Adam, "Country Review: Maldives," in C. De Young (ed.), *Review of the State of World Marine Capture Fisheries Management: Indian Ocean*, FAO Fisheries Technical Report No. 488 (Rome: FAO 2006). The map reflects the EEZ coordinates shown in the Maritime Zones of Maldives Act No.6/96 of June 1996, as notified to the United Nations Law of the Sea Secretariat (<http://www.un.org/Depts/los/LEGISLATIONANDTREATIES/PDFFILES/MDV_1996_Act.pdf>). While these coordinates had been discussed with the UK at technical level in November 1992, a draft agreement was never signed and is not in force; see the written answer by FCO Minister of State Dr. Kim Howells, *House of Commons Debates* vol. 470 (January 9, 2008), col. 559; and *cf.* p. 67.
51. While the print version of the CIA *World Factbook* (Langley/VA: U.S. Central Intelligence Agency 1995) correctly states (at p. 60) that "the entire Chagos Archipelago is claimed by Mauritius," the on-line version (as updated on May 15, 2008) asserted that "Mauritius and Seychelles claim the Chagos Archipelago including Diego Garcia." *Cf.* generally C.J. Harwood Jr., "CIA World Factbook (Diego Garcia) and Criminal Lies," <http://homepage.ntlworld.com/jksonc/5_cia-lies.html> (updated May 21, 2008).
52. See the comments on Diego Garcia by the former President of the Seychelles, Sir James R. Mancham (note 11) pp. 41–50; and by Mike O'Brien, UK Secretary of State for Foreign and Commonwealth Affairs, *House of Commons Debates* vol. 390 (October 15, 2002), col. 529W: "Seychelles makes no claim to sovereignty over the Chagos Islands."
53. See United Kingdom Territorial Sea Act 1987 (in force October 1, 1987), with Territorial Sea (Limits) Order 1989, Statutory Instruments [1989] No. 482 of March 15, 1989 (in force April 6, 1989); and US Presidential Proclamation No. 5928 of December 27, 1988, 54 Federal Register 777 (1988).
54. British Indian Ocean Territory Administration, *Laws and Guidance for Visitors* (London: Foreign and Commonwealth Office, February 2009), section 'Diego Garcia': "Unauthorised vessels or persons are not permitted to access this island and no unauthorised vessel is permitted to approach within 12 nautical miles."
55. See the 1991 and 2003 proclamations cited in note 46.
56. See Mauritius Governmental Notice No. 126 (Maritime Zones Regulations) of August 5, 2005, implementing the Maritime Zones Act No. 2 of February 28, 2005, and showing the geographical coordinates delimiting the exclusive economic zone (EEZ) of Mauritius as including the Chagos Archipelago—as already claimed in the earlier Maritime Zones Regulations No. 199 of December 1984 (table C1.T165).
57. The United States raised objections in 1982 against the earlier Maritime Zones Act No. 13 and Proclamation No. 7 of August 1977, which required foreign warships to give notification before transiting the territorial waters of Mauritius. On current use of Diego Garcia

NOTES—CHAPTER 1

by nuclear-powered warships and for the stationing of nuclear weapons, see pp. **39–40** and Chapter 3 (notes 41–49).

58. As confidently assumed by K.E. Calder, *Embattled Garrisons: Comparative Base Politics and American Globalism* (Princeton/NJ: Princeton University Press 2007) p. 231 ("one ideal location"). See also the Congressional Research Service Report to the US Senate Committee on Foreign Relations, *United States Foreign Policy Objectives and Overseas Military Installations*, 96th Congress, 1st Session (Washington D.C.: U.S. Government Printing Office, April 1979) p. 93, claiming that "Diego Garcia is immune to local political developments"; T.B. Millar, "Geopolitics and Military/Strategic Potential," in A.J. Cottrell and R.M. Burrell (eds.), *The Indian Ocean: Its Political, Economic and Military Importance* (New York: Praeger 1972) pp. 63–77, at p. 72 ("invulnerable to local political pressures, because there virtually are none"); A.J. Cottrell and Associates, *Sea Power and Strategy in the Indian Ocean* (Beverly Hills: Sage 1981) p. 125 ("relatively immune to political disturbance"); R.E. Harkavy, 'Thinking About Basing', in C. Lord (ed.), *Reposturing the Force: U.S. Overseas Presence in the Twenty-First Century* (Newport, RI: Naval War College Papers 2006) pp. 17-39, at 28 ("immune to future political threats"); and Erickson et al. (note 11) p. 27 ("does not depend on the acquiescence of capricious local governments"). But see T.P. Lynch, "Diego Garcia: Competing Claims to a Strategic Isle," *Case Western Reserve Journal of International Law* vol. 16 (1984) pp. 101–123; and J. Larus, "Diego Garcia: The Military and Legal Limitations of America's Pivotal Base in the Indian Ocean," in W.L. Dowdy and R.B. Trood (eds.), *The Indian Ocean: Perspectives on a Strategic Area* (Durham/NC, Duke University Press 1985) pp. 435–451, at p. 436.

59. On the implications for military transit rights see [Cdr.] S.A. Rose, "Operational Law: Naval Activity in the EEZ—Troubled Waters Ahead?," *Naval Law Review* vol. 39 (1990) pp. 67–92; J.M. Van Dyke, "The Disappearing Right to Navigational Freedom in the Exclusive Economic Zone," *Marine Policy* vol. 29 (2005) pp. 107–121; M. Valencia and K. Akimoto, "Guidelines for Navigation and Overflight in the Exclusive Economic Zone," *Marine Policy* vol. 30 (2006), pp. 704–711; and [Comm.] S. Bateman, "The Regime of the Exclusive Economic Zone: Military Activities and the Need for Compromise?," in T.M. Ndiaye and R. Wolfrum (eds.), *Law of the Sea, Environmental Law and Settlement of Disputes: Liber Amicorum Judge Thomas A. Mensah* (Leiden: Nijhoff 2007) pp. 569–582.

60. Geneva Conventions III and IV of August 12, 1949 (in force October 21, 1950), 75 United Nations Treaty Series 135 and 287; ratification by the UK on September 23, 1957 (*without* extension to overseas territories), accession by Mauritius on August 18, 1970; see also p. **64**.

61. Geneva Convention, Protocols I (International Armed Conflicts) and II (Non-International Armed Conflicts) of June 8, 1977, 1125 United Nations Treaty Series 3 and 609, ratified by Mauritius on March 22, 1982, and by the United Kingdom on January 28, 1998, with an extension to BIOT on July 2, 2002; see also the Geneva Convention (Amendment) Act (Overseas Territories) Order 2002, of April 17, 2002 (effective May 1, 2002), Statutory Instruments [2002] No. 1076. Mauritius formally objected to the BIOT extension on June 27, 2003, asserting its sovereignty over the Chagos Archipelago including Diego Garcia.

62. UN Covenants on Human Rights (993 United Nations Treaty Series 3, and 999 United Nations Treaty Series 171); accession by Mauritius on December 12, 1973, ratification by the United Kingdom on May 20, 1976.

63. UN Convention Against Torture and Other Cruel, Inhuman or Degrading Treatment or Punishment (1465 United Nations Treaty Series 85), accession by Mauritius on December 9, 1992, ratification by the United Kingdom on December 8, 1988 (with an extension to British dependent territories *except for BIOT*, declared on December 9, 1992); see also Chapter 4 (note 22). Similarly, UK ratification (on June 24, 1988) of the 1987 European Convention for the Prevention of Torture and Inhumane or Degrading Treatment or Punishment (1561 United Nations Treaty Series 363) was extended to Gibraltar and Guernsey (on September 5, 1988 and November 8, 1994, respectively), but *not* to BIOT.

64. The UK Foreign and Commonwealth Office contends that its ratification of the 1966 UN Covenants on Human Rights (note 62) does *not* extend to BIOT; see H. Fox, "United Kingdom of Great Britain and Northern Ireland: Dependent Territories," in R. Bernhardt (ed.), *Encyclopedia of Public International Law* vol. 4 (Amsterdam: Elsevier 2000) pp. 1025–1029, at p. 1029; see also Chapter 2 (note 18). However, the UN Human Rights Committee has repeatedly indicated that it considers the covenants to apply to BIOT, and in its concluding observations on the UK report to the 2008 meeting urged the United Kingdom "to include the territory in its next periodic report"; see the Report of the Committee on its 93rd Session (Geneva, July 7–25, 2008), UN-Doc. CCPR/C/GBR/CO/6 (July 30, 2008) p. 6/paragraph 22, and *cf.* the submission dated October 12, 2007 from the [non-governmental] Minority Rights Group International to the House of Commons Select Committee on Foreign Affairs, *Overseas Territories: Seventh Report of Session 2007–08*, vol. II, HC 147-II (London: HMSO, July 6, 2008) pp. Ev115–125, at p. 116. See also S. Allen, "International Law and the Resettlement of the (Outer) Chagos Islands," *Human Rights Law Review* vol. 8 (2008) pp. 683–702; and Chapter 6 (notes 13, 20, 33). On the European Convention on Human Rights, see Chapter 2 (note 74).
65. Statute of the International Criminal Court (Rome, July 17, 1998, in force July 1, 2002), 2187 United Nations Treaty Series 3, ratified by the United Kingdom on October 4, 2001 (*without* extension to overseas territories), and by Mauritius on March 5, 2002. The United States signed the Rome Statute on December 31, 2000, but (on May 6, 2002) notified the UN Secretariat that it "does not intend to become a party to the treaty" and accordingly "has no legal obligations arising from its signature."
66. See U.S. Presidential Determination No. 2003-28 of July 29, 2003, pursuant to the American Servicemen's Protection Act of 2002 (title II of Public Law 107-206), 22 U.S. Code 7421 [note, however, that the bilateral immunity agreement has still *not* been ratified by the Mauritian Parliament]; see also H. van der Wilt, "Bilateral Agreements between the United States and States Parties to the Rome Statute: Are They Compatible with the Object and Purpose of the Statute?," *Leiden Journal of International Law* vol. 18 (2005) pp. 93–111; and J. Kelley, 'Who Keeps International Commitments and Why? The International Criminal Court and Bilateral Nonsurrender Agreements', *American Political Science Review* vol. 101 (2007) pp. 573–589.
67. On "black hole" theories in the U.S., see P.B. Heyman, *Terrorism, Freedom, and Security: Winning Without War* (Cambridge/MA: MIT Press 2003) p. 162; J. Steyn, "Guantánamo Bay: The Legal Black Hole," *International and Comparative Law Quarterly* vol. 53 (2004) pp. 1–15; R. Wilde, "Legal 'Black Hole'? Extraterritorial State Action and International Treaty Law on Civil and Political Rights," *Michigan Journal of International Law* vol. 26 (2005) pp. 739–806, at pp. 744 and 749; P. Sands, *Lawless World: Making and Breaking Global Rules* (London: Penguin Books 2006) pp. 143–173; N.R. Sonnett, "Guantanamo: Still a Legal Black Hole," *Human Rights* vol. 33 No. 1 (2006) pp. 8–9; and J. Mayer, "The Black Sites," *New Yorker* vol. 83 No. 24 (August 13, 2007).—On "human rights black holes" in UK overseas territories, see p. **64**, and Chapter 2 (notes 69–74).
68. Appendix I, Annex II, Art. 1(b)(ii) [p. **76**, emphasis supplied]. As pointed out by Woodliffe (note 27) at p. 173, this provision is modelled after Art. VII(2)(a) of the 1951 NATO Status of Forces Agreement (SOFA), 199 United Nations Treaty Series 67; see [Lt. Col.] J.H. Rouse and [1st Lt.] G.B. Baldwin, "The Exercise of Criminal Jurisdiction Under the NATO Status of Forces Agreement," *American Journal of International Law* vol. 51 (1957) pp. 29–62. The "law in force in the territory" was subsequently defined (by section 3 of the BIOT Courts Ordinance of 1983) as the law of England, "provided that the said law of England shall apply in the Territory only so far as it is applicable and suitable to local circumstances and shall be construed with such modifications, adaptations, qualifications and exceptions as local circumstances render necessary"; see also p.64.

NOTES—CHAPTER 1

69. See the *Landmine Monitor Report 1999: Toward a Mine-Free World*, by the NGO International Campaign to Ban Landmines (ICBL, co-laureate of the 1997 Nobel Peace Prize) pp. 328–334, citing U.S. sources for Diego Garcia stocks as of 1997; see also C.W. Jacobs [Maj., US Army Judge Advocate], "Taking the Next Step: An Analysis of the Effects the Ottawa Convention May Have on the Interoperability of United States Forces with the Armed Forces of Australia, Great Britain and Canada," *Military Law Review* vol. 180 (2004) pp. 49–114, at p. 67. *Cf.* U.S. General Accounting Office, *Military Operations: Information on U.S. Use of Land Mines in the Persian Gulf War*, GAO-02-1003 (Washington, D.C.: GAO, September 30, 2002) p. 39 (table 7, showing a total of at least 10.3 million anti-personnel landmines stockpiled by the U.S. Department of Defense).
70. Convention on the Prohibition of the Use, Stockpiling, Production and Transfer of Anti-Personnel Mines and on their Destruction, adopted at Ottawa on September 18, 1997 (in force March 1, 1999), text in 2056 United Nations Treaty Series 211; ratified by Mauritius on December 3, 1997, and by the United Kingdom (with explicit extension to BIOT) on July 31, 1998. See also J.R. Crook, "Contemporary Practice of the United States Relating to International Law: U.S. Policy Regarding Landmines," *American Journal of International Law* vol. 102 (2008) pp. 190–191. Even though section 5(5) of the UK Landmines Act of July 28, 1998 (Public Acts 1998, Chapter 33) exempts "international military operations" involving "members of the armed forces of a State that is not a party to [the Ottawa] Convention" from the prohibition of section 2 (possession and transfer of anti-personnel mines), that does *not* exonerate the UK from its annual reporting duties regarding the total of "all stockpiled mines...under its jurisdiction" pursuant to Art. 7(1)(b) of the Convention.
71. Written answer by the Secretary of State for Foreign and Commonwealth Affairs, *House of Commons Debates* vol. 345 (March 6, 2000), col. 504W [emphasis supplied]; see also the letter dated February 25, 2003 from Adam Ingram, UK Minister of State for the Armed Forces, to the *Diana Princess of Wales Memorial Fund* and the ICBL; quoted in Jacobs (note 69) p. 95, footnote 182.
72. Statement by Ambassador David Broucher, Ottawa Convention Standing Committee meeting on May 16, 2003, "General Status and Operation of the Convention—United Kingdom Intervention on Article 1," as quoted in Jacobs (note 69) p. 96. Yet, in the view of the International Committee for the Red Cross, "permitting the transit of antipersonnel mines through the territory of a State Party would undermine the object and purpose of the [Convention]...and contradict its prohibition on assisting anyone in the stockpiling and use of antipersonnel mines"; see ICBL Report 1999 (note 69) Annex, pp. 1005–1006. *Cf.* S. Maslen, *Commentaries on Arms Control Treaties: The Convention on the Prohibition of the Use, Stockpiling, Production, and Transfer of Anti-Personnel Mines and on Their Destruction* (Oxford: Oxford University Press 2004) p. 100.
73. Convention on the Prohibition of the Development, Production, Stockpiling and Use of Chemical Weapons and on their Destruction, adopted on September 3, 1992 (in force April 29, 1997), text in 1974 United Nations Treaty Series 45; ratified by Mauritius on February 9, 1993, by the United Kingdom on May 13, 1996 (ratification extended to BIOT on October 26, 2005), and by the United States on April 25, 1997; see note 45.
74. Convention on Cluster Munitions, adopted at Dublin on May 30, 2008, and signed at Oslo on December 3, 2008, by 94 countries (including the UK, though not the USA), text at <http://www.clustermunitionsdublin.ie/pdf/ENGLISHfinaltext.pdf>; for background see V. Wiebe, "Footprints of Death: Cluster Bombs as Indiscriminate Weapons Under International Humanitarian Law," *Michigan Journal of International Law* vol. 22 (2000) pp. 85–168; and P. Pillai, "Adoption of the Convention on Cluster Munitions," *American Society of International Law: Insights* vol. 12 No. 20 (online October 1, 2008). Art. 21(3), built into the Convention at the initiative of Germany and other NATO countries concerned

about their military "inter-operative" relationship with the United States, allows a member State to "engage in military cooperation and operations with States not party to this Convention that might engage in activities prohibited to a State Party"—similar to section 5(5) of the UK Landmines Act (note 70). That will *not*, however, exonerate it from its annual reporting duties under Art. 7(1)(b) of both conventions for the total of all prohibited munitions "under its jurisdiction and control"; *cf.* Maslen (note 72) p. 203.

75. Map shown as figure B-1 in the Navy's 1997 *Natural Resources Management Plan Diego Garcia* (Chapter 5, note 41) p. B-4; "redacted" copy made available to the author on July 19, 2007, by direction of the Commander, U.S. Naval Forces Japan, under FOIA request CNFJ 07-26. U.S. Navy safety standards require a minimum distance between vessels used for ship-to-ship transfers of ordnance, hence the wide circular "ESQD" (= Explosive Safety Quantity Distance) perimeters shown on the map.
76. See also p. **48**. On ship-based ammunition storage at Diego Garcia, see E.J. Labs, *The Future of the Navy's Amphibious and Maritime Prepositioning Forces* (Washington D.C.: Congressional Budget Office, November 2004) p. 6.
77. Letter dated November 19, 2003 from Peter Carter QC, chair of the BHRC of England and Wales, to foreign secretary Jack Straw; as quoted in Jacobs (note 69) p. 95; cf. *Diego Garcia: Footprint of Freedom?* (London: Bar Human Rights Committee, November 17, 2003). It is noteworthy that *Norway*, which also hosts a U.S. military base, successfully insisted on complete removal of all prohibited ammunition by March 1, 2003, in order to comply with the Ottawa Convention; ICBL Landmine Monitor Report (note 69) 2001.

2 Human Rights: How to Depopulate an Island

1. P. Leymarie, "Grandes manoeuvres dans l'Océan Indien: la base de Diego-Garcia, sur la route des pétroliers et des cargos," *Le Monde Diplomatique* vol. 23 (December 1976) p. 19, noting that "aircraft taking off from Diego Garcia do not risk cyclones or heavy storms that are frequent in other parts of the ocean."
2. See Chapter 1: Scott (note 2), Edis (note 3) pp. 29–52, and Pilger (note 7) p. 20.
3. On the history of slavery in Diego Garcia, see Vine (Chapter 1, note 6) pp. 60–69, and D. Taylor, "Slavery in the Chagos Archipelago," *Chagos News* No. 14 (2000) pp. 2–4.
4. I owe this information to my colleague, veteran Diego Garcia "wizard" Steven J. Forsberg (Chapter 3, note 36), who is currently at work on a historical study of the topic. See also G. Powell, *The Kandyan Wars: The British Army in Ceylon 1803–1818* (London: Cooper 1973) pp. 148 and 181; S. de Silva Jayasuriya, "Trading on a Thalassic Network: African Migrations Across the Indian Ocean," *International Social Science Journal* vol. 58 (2006) pp. 215–225, at p. 219; *id.*, "Afro-Sri Lankais: liens et raciness," *Cahiers des Anneaux de la Mémoire: Europe, Afrique, Amériques* No. 9 (2006) pp. 171–186, at p. 179; and *id.*, "Migrants and Mercenaries: Sri Lanka's Hidden Africans," in S. de Silva Jayasuriya and J.P. Angenot (eds.), *Uncovering the History of Africans in Asia* (Leiden: Brill 2008) pp. 155–170, at p. 159.
5. C. Pridham, *England's Colonial Empire: Mauritius and its Dependencies* (London: Elder, Smith & Co. 1846) p. 403, and Edis (Chapter 1, note 3) p. 39.
6. See Chapter 1 (note 14).
7. As quoted by Lord Justice Laws in *The Queen (ex parte Bancoult) vs. Foreign Secretary (DC)*, Chapter 1 (note 16) p. 1082.
8. *Ibid.* at p. 1083, with another official's handwritten comment on this note, facetiously referring to "a certain old-fashioned reluctance to tell a whopping fib, or even a little fib, depending on the number of permanent inhabitants."
9. As quoted by Laws L.J., *ibid.*, who labeled these documents "embarrassing or worse" (at p. 1106). The minute of August 24, 1966, was signed by the permanent under-secretary's

private secretary, Patrick R.H. (later Sir Patrick) Wright. There is no reference to BIOT in the memoirs of P. Gore-Booth, *With Great Truth and Respect* (London: Constable 1974).
10. BIOT Ordinance No. 1 of February 8, 1967 (*Compulsory Acquisition of Land for Public Purposes*), and No. 2 of March 22, 1967 (*Acquisition of Land for Public Purposes: Private Treaty*); both repealed by a 1983 ordinance that declared all BIOT land Crown property. According to a Mauritian parliamentarian, however, legal doubts remain with regard to the acquisition of freehold on some of the islands, in view of recently discovered British royal concessions dating back to 1865; see "Héritage spolié? La famille Lemière revendique huit îles de Chagos," *Le Mauricien* (Port Louis, November 11, 2007, online).
11. See G. M.F. Drower, *Britain's Dependent Territories : A Fistful of Islands* (Aldershot : Dartmouth 1992) p. 149; see also Chapter 1 (note 19).
12. The company's copra production on the outer Chagos islands of Peros Banhos and Salomon, about 135 miles [218 kilometers] north of Diego Garcia, also terminated in 1972.
13. Described in chronological detail by Bezboruah (Chapter 1, note 6) pp. 57–66 and Bandjunis (Chapter 1, note 6) pp. 35–43; see also Bhatt (Chapter 1, note 11) p. 30.
14. Letter from Eleanor J. Emery (head of the Pacific and Indian Ocean Department, later British ambassador to South Africa) to Sir Bruce Greatbatch (governor of the Seychelles), HPN 10/1; as quoted by T. Dalyell MP, *House of Commons Debates* vol. 360 (January 9, 2001) col. 181–184. See also M. Parris, "You are never too old to face your lies," *Times* (London, December 16, 2000) p. 24, and T. Slessor, *Ministries of Deception: Cover-Ups in Whitehall* (London: Aurum Press 2002) p. 22.
15. Classified memorandum from Jonathan D. Stoddart (Director, Office of International Security Operations, U.S. Department of State) dated December 16, 1970, as quoted by Pilger (Chapter 1, note 7) pp. 25, 46, and 317/footnote 6. For an account of the instrumental role of the U.S. military—including Admirals Paul Nitze and Elmo Zumwalt in particular—in the "relocation" of the islanders, see Vine (Chapter 1, note 6) pp. 168–193, quoting Admiral Zumwalt's unequivocal comment ("absolutely must go"); Office of the Chief of Naval Operations, CNO Command Sheet 1081-71, 'Re: Copra Workers in Diego Garcia' (March 26, 1971). See also D.R. Snoxell, 'Anglo/American Complicity in the Removal of the Inhabitants of the Chagos Islands', *Journal of Imperial and Commonwealth History* vol. 37 (2009) pp. 127–134.
16. Memorandum dated January 26, 1971, from Ian Watt, UK Foreign and Commonwealth Office (Atlantic and Indian Ocean Department), to D.A. Scott [(Sir) David Aubrey Scott, assistant private secretary to the secretary of state, FCO, later British high commissioner to New Zealand and governor of Pitcairn Islands] *et al.*, circulated on February 2, 1971; certified copy filed on April 22, 2002, with the U.S. District Court for the District of Columbia, as supplementary evidence in *Olivier Bancoult et al. vs. Robert S. McNamara et al.* (notes 38 and 67).
17. Bandjunis (Chapter 1, note 6) pp. 47–48.
18. Adopted at Strasbourg on September 16, 1963 (in force May 2, 1968), and amended on November 1, 1998, 1496 United Nations Treaty Series 263; see J.E.S. Fawcett, *The Application of the European Convention on Human Rights* (2nd edn. Oxford: Clarendon 1987) p. 422. Article 3 of the protocol provides that "(1) no one shall be expelled, by means either of an individual or of a collective measure, from the territory of the State of which he is a national," and "(2) no one shall be deprived of the right to enter the territory of the State of which he is a national." Article 12(4) of the 1966 International Covenant on Civil and Political Rights (see Chapter 1, notes 62 and 64) similarly declares that "no one shall be arbitrarily deprived of the right to enter his own country"; however, the United Kingdom upon its ratification of the covenant on May 20, 1976, expressly reserved the right *not* to apply Article 12(4) with regard to immigration laws for its dependent territories. See G. Goodwin-Gill, "The Limits of the Power of Expulsion in Public International Law," *British Yearbook of International Law* vol. 47 (1975) pp. 55–156, at pp. 140–41; R. Plender,

International Migration Law (2nd edn. Dordrecht: Nijhoff 1988) pp. 133–135, 142/footnote 128; and A.W.B. Simpson, *Human Rights and the End of Empire: Britain and the Genesis of the European Convention* (Oxford: Oxford University Press 2001) at pp. 1097–1098.

19. Secret minute dated October 23, 1968, as quoted by Laws L.J. (Chapter 1, note 16) p. 1085 and Pilger (Chapter 1, note 7) pp. 30, 38, and 317/footnote 11; see also generally A.I. Aust, *Handbook of International Law* (Cambridge: Cambridge University Press 2005) p. 40/footnote 33; Aust (note 73); and S. Jones, 'Indian Ocean Belongers, 1668–2008', *Southern Perspectives* (University of Technology Sydney, on-line March 24, 2009).

20. Quoted by Laws L.J. *ibid*. p. 1086.

21. Lady Hazel Fox CMG QC (Chapter 1, note 64), at p. 1026, cites BIOT among examples of "those territories which, by reason of the absence of any permanent population, do not satisfy the criteria of effective statehood" for the purposes of Article 73 (Chapter 1, note 14); see also Chapter 6 (note 25).

22. Quoted by Laws L.J. (Chapter 1, note 16) p. 1086; see also Slessor (note 14) p. 21.

23. M. Curtis, *Web of Deceit: Britain's Role in the World* (London: Vintage 2003) p. 421; *id*., *Unpeople: Britain's Secret Human Rights Abuses* (London: Vintage 2004).

24. Edis (Chapter 1, note 3) pp. 84–85. The majority of the inhabitants of Diego Garcia were provisionally moved to the Peros Banhos and Salomon islands, but all ended up eventually in Mauritius or the Seychelles.*Cf.* J.M.G. Le Clézio (Nobel laureate 2008), 'Archipel des Chagos—Diego Garcia: les déportés du paradis', *Le Point* No. 1578 (January 19, 2007).

25. See Chapter 1: Winchester (note 16) pp. 207–208, Pilger (note 7) pp. 26–28, 35, and Vine (note 6) p. 207.

26. *Diego Garcia, 1975: The Debate over the Base and the Island's Former Inhabitants*: Hearings, House of Representatives, 94th Congress, 1st Session, June 5 and November 4, 1975 (Washington, D.C.: U.S. Government Printing Office 1975); see also Bezboruah (Chapter 1, note 6) pp. 92–127. At the request of Senators John Culver and Edward Kennedy (Amendment 884 to Senate Bill 1517), the White House had to provide a Report on the Resettlement of Inhabitants in the Chagos Archipelago (October 10, 1975, drafted by George T. Churchill, director of the Office of International Security Operations, U.S. Department of State, and Cdr. Gary S. Sick, director for the Persian Gulf and Indian Ocean in the Office of International Security, U.S. Department of Defense).

27. As quoted by Bandjunis (Chapter 1, note 6) p. 128. Similar statements, referring to British assurances that there was no resident population, were made by retired U.S. Admiral Elmo R. Zumwalt and by Seymour Weiss, director of the State Department's Bureau of Politico–Military Affairs (testifying before the House of Representatives' Foreign Affairs Sub-Committee on the Near East and South Asia on March 6, 1974: "no local population whatsoever"), reprinted in Jawatkar (Chapter 1, note 28) p. 314. See also F.A. Váli, *Politics of the Indian Ocean Region: The Balances of Power* (New York: Free Press 1976) p. 202 ("non-permanent" inhabitants "from the Seychelles, Mauritius, and other islands"); but *cf.* the critical comments by Senator John C. Culver, U.S. *Congressional Records*, Senate, 94th Congress, 1st Session, vol. 121 No. 133 (11 September 1975) pp. S15865–15868; and Calder (Chapter 1, note 58) p. 187 (Diego Garcia "finessed" as a "rent-a-rock" platform for global power projection).

28. D.B. Ottaway, "Islanders Were Evicted for U.S. Base," *Washington Post* (September 9, 1975) p. A1; and editorial "The Diego Garcians," *Washington Post* (September 11,1975).

29. "Expulsion" has been defined as "an act, or a failure to act, by a State with the intended effect of securing the departure of persons against their will from the territory of that State"; L.T. Lee, "Draft Declaration of Principles of International Law on Mass Expulsion," *Proceedings of the American Society of International Law* vol. 78 (1984) pp. 344–346. According to L.B. Sohn and T. Buergenthal, *The Movement of Persons across Borders* (Washington, D.C.: American Society of International Law 1992) at p. 85, "it is a firmly established rule of international law that a State may not deport or expel its own nationals." See also the International

Notes—Chapter 2

Law Association's Declaration of Principles of International Law on Mass Expulsion (65th Conference Report, Cairo 1992); A.M. de Zayas, "Population, Expulsion and Transfer," in R. Bernhardt (ed.), *Encyclopedia of Public International Law* vol. 3 (Amsterdam: Elsevier 1997) pp. 1062–1068, at pp. 1063–1064; and J.M. Henckaerts, *Mass Expulsion in Modern International Law and Practice* (The Hague: Martinus Nijhoff 1995) p. 80, with a "representative list of mass expulsions since 1945" on pp. 204–205 [to which the Chagos Islands could rightfully be added].

30. George T. Churchill (see note 26), as quoted by Bandjunis (Chapter 1, note 6) p. 132; see also note 40.
31. See J. Madeley, *Diego Garcia: A Contrast to the Falklands* (London: Minority Rights Group 1985) pp. 5–7, also pointing out that "slowness on the part of the Mauritian Government added to the exiles' problems"; confirmed by Oraison (Chapter 1, note 18) p. 175.
32. "Diego Garcia: The Islanders Britain Sold," *Sunday Times* (London, September 21, 1975), reprinted in the report of the U.S. Congressional Hearings (note 26) pp. 93–101; H. Siophe, "Summary of Survey of the Persons Displaced from the Chagos Islands," *ibid.* pp. 112–121; "La grande misère des déportés de Diego–Garcia," *Le Monde* (Paris, September 26, 1975); see also C. Zorgbibe, "L'affaire Diego Garcia," *Le Monde Diplomatique* vol. 27 No. 314 (May 1980) p. 18. A subsequent ethnographic study found that even a decade later, the exiled Îlois were still viewed as alien elements by indigenous Mauritians and suffered from heavy discrimination; see I.B. Walker, *Zaffer pe Sanze: Ethnic Identity and Social Change Among the Îlois in Mauritius* (Vacoas: KMLI 1986) pp. 21–24.
33. A.R.G. Prosser, *Mauritius: Resettlement of Persons Transferred from Chagos Archipelago* (Port Louis: Government Printer 1976) p. 6. See also H. Sylva, *Report on the Survey on the Conditions of Living of the Îlois Community Displaced from the Chagos Archipelago* (Port Louis: Ministry of Social Security, April 22, 1981); D. Vine, "War and Forced Migration in the Indian Ocean: The US Military Base at Diego Garcia," *International Migration* vol. 42 (2004) pp. 111–143; *id.*, Chapter 1 (note 6) pp. 212–257; S.F. Johannessen, Contested Roots: The Contemporary Exile of the Chagossian Community in Mauritius (MA Thesis: University of Oslo 2005); L. Jeffery, The Politics of Victimhood Among Displaced Chagossians in Mauritius (PhD Dissertation: Cambridge University 2006); and the novels by Peter Benson, *A Lesser Dependency* (London: Macmillan 1989; BBC radio play 1991), and Shenar Patel, *Le silence des Chagos* (Paris: Editions de l'Olivier 2005). A stageplay on Diego Garcia by Robert B. Sherman was considered too contentious to be performed in London; see his obituary, *Times* (London, September 11, 2004), p. 44.
34. Footnote 3 in the *Deed of Acceptance and Power of Attorney*, annexed to Madeley (note 31) p. 15.
35. Collected in Port Louis in November 1979 by a prominent British human rights solicitor, Bernard Sheridan; see Madeley, *ibid.* p. 8, and Slessor (note 14) p. 26. For a detailed chronological description of the negotiations, see the 2003 high court decision (note 48).
36. Madeley (note 31) p. 11; see also Oraison (Chapter 1, note 18) p. 176.
37. Signed by Mauritian foreign minister Jean Claude de l'Estrac and the UK high commissioner to Mauritius, James Nicholas Allan [CMG CBE, later British ambassador to Mozambique], entered into force pursuant to an exchange of notes on October 26, 1982 (Cmnd. 8789), registered with the UN Secretariat by the UK Foreign and Commonwealth Office on May 31, 1983; text in 1316 United Nations Treaty Series 127 (No. 21924), see also *British Yearbook of International Law* vol. 53 (1982) p. 489.
38. *Olivier Bancoult et al. vs. Robert S. McNamara et al.*, in the U.S. District Court for the District of Columbia; see 227 *Federal Supplement* 2nd Series 144 (2002), 214 *Federal Rules Decisions* 5 (2003), 217 *Federal Rules Decisions* 280 (2003), and 370 *Federal Supplement* 2nd Series 1 (D.D.C. 2004). For background of claims, see D. Vine *et al.*, *Dérasiné: The Expulsion and Impoverishment of the Chagossian People (Diego Garcia)*, Expert Report prepared for American University Washington College of Law and Sheridan Solicitors (London, April 11, 2005); C. Nauvel,

"A Return from Exile in Sight? The Chagossians and Their Struggle," *Northwestern Journal of International Human Rights* vol. 5 (2006) pp. 96–126; and P. Harvey *et al.*, "'We All Must Have the Same Treatment': Calculating the Damages of Human Rights Abuses for the People of Diego Garcia," in B.R. Johnston and S. Slyomovics (eds.), *Waging War, Making Peace: Reparations and Human Rights* (Tucson/AZ: Left Coast Press 2008). The Alien Tort Claims Act (28 U.S. Code 1350, originally part of the Judiciary Act of 1789) permits federal courts to hear claims by aliens for torts committed "in violation of the law of nations or a treaty of the United States"; see A. Ridenour, "Apples and Oranges: Why Courts Should Use International Standards to Determine Liability for Violations of the Law of Nations Under the Alien Tort Claims Act," *Tulane Journal of International and Comparative Law* vol. 9 (2001) pp. 581–603, and L. Westra, *Environmental Justice and the Rights of Indigenous Peoples: International and Domestic Legal Perspectives* (London: Earthscan 2008) at pp. 109–111, 167–168.
39. *De Chazal du Mée vs. Olivier Bancoult* et al., Supreme Court of Mauritius, judgment of August 7, 2002; text filed on August 19, 2002, as evidence for case 01-2619, D.C. District Court (note 38).
40. *Olivier Bancoult et al. vs. Robert S. McNamara et al.* (April 21, 2006), 445 *Federal Reporter* 3rd Series 427 (D.C. Cir. 2006); see the note by J.R. Crook, "Contemporary Practice of the United States Relating to International Law," *American Journal of International Law* vol. 100 (2006) p. 692. The claim against *Halliburton Corp.* was dropped by the plaintiffs in 2003, and statutory immunity was granted under the Federal Employees Liability Reform and Tort Compensation Act of 1988 (Westfall Act)—Public Law 100–694, 102 *Statutes at Large* 4563 (1988), 28 *United States Code* 2671—also to the two State Department officials cited as defendants, George T. Churchill (note 26) and Eric Newsom (note 46); see Westra (note 38) at p. 109. For criticism see K. Lin, "An Unintended Double Standard of Liability: The Effect of the Westfall Act on the Alien Tort Claims Act," *Columbia Law Review* vol. 108 (2008) pp. 1718–1757 ("impassable barrier to ATCA claims").
41. 127 *Supreme Court Reporter* 1225 (2007).
42. D.C. Circuit Court of Appeals (opinion for the court filed by Judge Brown, note 40) at pp. 435, 437; and Crook (note 40) at p. 693. On the doctrinal debate see F.W. Scharpf, "Judicial Review and the Political Question: A Functional Analysis," *Yale Law Journal* vol. 75 (1966) pp. 517–597; T. Franck, *Political Questions/Judicial Answers: Does the Rule of Law Apply to Foreign Affairs?* (Princeton/NJ: Princeton University Press 1992); and J.H. Choper, "The Political Question Doctrine: Suggested Criteria," *Duke Law Journal* vol. 54 (2005) pp. 1457–1524, at p. 1465 (arguing in favour of judicial review to protect individual rights).
43. *Cf.* the note on *Olivier Bancoult vs. Robert S. McNamara et al.*, *Harvard Law Review* vol. 120 (2007) pp. 860–867; and R.H. Fallon, "The Core of an Uneasy Case for Judicial Review," *Harvard Law Review* vol. 121 (2008) pp. 1694–1736. However, the District of Columbia Circuit Court confirmed its position in *Jennifer Harbury vs. Michael V. Hayden et al.* (D.C.Cir., April 15, 2008), 522 *Federal Reporter* 3rd Series 413, citing *Bancoult vs. McNamara* with approval.
44. After initial refusals by the UK authorities, as reported in *British Yearbook of International Law* vol. 70 (1999) pp. 456–457, the British Overseas Territories Act of 2002 confirmed the Chagossians' full entitlement to citizenship rights; see *British Yearbook of International Law* vol. 73 (2002) p. 593. The House of Commons Foreign Affairs Committee recommended to extend citizenship also "to third generation descendants of exiled Chagossians"; *Overseas Territories*, 7th report of session 2007–08, HC 147-I (London: HMSO, July 6, 2008) pp. 35 and 138. However, in the context of the 2009 Borders, Citizenship and Immigration Bill, the House of Lords deferred the 'sensitive issue' of the Chagossians to further discussions with the Foreign Office; *House of Lords Debates* vol. 709 (April 1, 2009) col. 1040.
45. See S. Bangaroo, "A Short Analysis of the Exile of an Indigenous Population from Beginning to End," *Hertfordshire Law Journal* vol. 3 (2005) pp. 3–7, at p. 4; and the summary

of follow-up proceedings by S. Allen, "Responsibility and Redress: Chagossian Claims in the English Courts," in S.J.T.M. Evers (ed.), *The Fate of the Chagossians Since Their Eviction from the Chagos Islands* (Leiden: Brill, forthcoming 2009).

46. Letter from Eric Newsom, U.S. Assistant Secretary of State for Political–Military Affairs [later a partner in the Arlington/VA firm of *Collins & Co.*, specialized in "dual use" arms export licensing and lobbying; see also note 40], to Richard Wilkinson CVO, director for the Americas Command in the UK Foreign and Commonwealth Office [later British ambassador to Venezuela and Chile]; excerpts quoted in "US Blocks Return Home for Exiled Islanders," *Guardian* (London, September 1, 2000). On U.S. objections to resettlement see also Moor and Simpson (Chapter 1, note 23) p. 162 footnote 265, and Chapter 6 (note 9).

47. [2001] *Queen's Bench Reports* p. 1967, [2001] 2 *Weekly Law Reports* 1219, [2000] *All England Law Reports* (D) 1675, *International Law Reports* vol. 123 (2003) p. 555; case notes by M. Byers, *British Yearbook of International Law* vol. 71 (2000) p. 433, S. Palmer, "'They Made a Desert and Called It Peace': Banishment and the Royal Prerogative," *Cambridge Law Journal* vol. 60 (2001) pp. 234–236; and A. Tomkins, "Magna Carta, Crown and Colonies," [2001] *Public Law* pp. 571–585. See also L. Balmond, "Chronique des faits internationaux: Grande Bretagne/Îles Chagos," *Revue Générale de Droit International Public* vol. 104 (2000) p. 186; E. MacAskill and R. Evans, "Scandal of Diego Garcia: Thirty Years of Lies, Deceit and Trickery That Robbed a People of Their Island Home," *Guardian* (London, 4 November 2000) p. 3; H. Malaisé, "Exil forcé loin de Diego Garcia," *Le Monde Diplomatique* vol. 48 No. 573 (December 2001) p. 21; and Chapter 1 (note 16).

48. [2003] *England and Wales High Court Reports* (Queen's Bench) 2222, [2003] *All England Law Reports* (D) 166; case note by R. O'Keefe, *British Yearbook of International Law* vol. 74 (2003) pp. 486–493; see also L. Jeffery, "Historical Narrative and Legal Evidence: Judging Chagossians' High Court Testimony," *Political and Legal Anthropology Review* vol. 29 (2006) pp. 228–253.

49. [2004] *England and Wales Court of Appeal Reports* (Civ.) 997.

50. Excerpts in *British Yearbook of International Law* vol. 75 (2004) pp. 664–665.

51. Foreign and Commonwealth Office, written statement in the House of Commons Foreign Affairs Committee (7 December 2005), 1583–I; *British Yearbook of International Law* vol. 77 (2006) p. 639.

52. Statement by Bill Rammell, parliamentary under-secretary of state for foreign and commonwealth affairs (on July 7, 2004), in the House of Commons, answering the question as to "whether there was any pressure from the United States of America": "I have certainly received no representations from the United States and I do not believe that the Foreign Secretary has had any representations on the issue for a significant number of years"; *House of Commons Debates* vol. 423, col. 293WH, *British Yearbook of International Law* vol. 75 (2004) p. 670.

53. Letter dated November 16, 2004 by Lincoln P. Bloomfield Jr., Assistant Secretary of State for Political–Military Affairs [former Director of Global Issues in the U.S. National Security Council, and currently the State Department's Special Envoy for the Control of Man-Portable Air Defense Systems], to Robert N. Culshaw MVO, FCO Director of the Americas and Overseas Territories [later deputy director of the British Antarctic Survey], quoted by Lord Justice Hooper in *The Queen (ex parte Bancoult) vs. Foreign and Commonwealth Office* (note 62), at para. 96. But see Cooper Chapter 6 (note 9).

54. Hence *not* a "non-self-governing territory" subject to reporting duties under Art. 73 of the UN Charter (Chapter 1, note 14); Fox (note 21) p. 1026.

55. Reissued by the BIOT Administration as *Laws and Guidance for Visitors* (London: Foreign and Commonwealth Office, February 2009); see Chapter 1 (note 54). The new mooring fee had originally been set at £500 ($740), but was reduced after vehement protests by yacht-owners. New mooring sites for the 20–50 yachts visiting annually were designated by the

BIOT Administration in 2007, off the atolls of Salomon, Peros Banhos and Boddan [about 135 miles = 218 kilometers north of Diego Garcia]. See also H. Cloudview, "Chagos: An Unspoilt Paradise," *Villa and Yacht Magazine* vol. 2 No. 1 (2007) pp. 20–24; and K. Hubert, "The Chagos Archipelago: Paradise Lost?" *Sea Yachting* vol. 2 No. 6 (2007) pp. 35–41.
56. See Chapter 1 (notes 46 and 50).
57. See p. **17**, and Chapter 1 (note 50); *cf.* D. Bowett, "Islands," in R. Bernhardt (ed.), *Encyclopedia of Public International Law* vol. 2 (Amsterdam: Elsevier 1999) pp. 1455–1457. During UNCLOS negotiations, the United Kingdom had opposed Article 121(3) but was outvoted; see J.M. Van Dyke and R.A. Brooks, "Uninhabited Islands: Their Impact on the Ownership of the Oceans' Resources," *Ocean Development and International Law* vol. 12 (1983) pp. 265–300; J. I. Charney, "Rocks That Cannot Sustain Human Habitation," *American Journal of International Law* vol. 93 (1999) pp. 863–878; and R. Lavalle, "Not Quite a Sure Thing: The Maritime Areas of Rocks and Low–Tide Elevations Under the U.N. Law of the Sea Convention," *International Journal of Marine and Coastal Law* vol. 19 (2004) pp. 43–70.
58. E.D. Brown, *The International Law of the Sea*, vol. 1 (Aldershot: Dartmouth 1994) p. 151. See also H. Dipla, *Le régime juridique des îles dans le droit international de la mer* (Paris: Presses Universitaires de France 1994) p. 41; B. Kwiatkowska and A.H.A. Soons, "Entitlement to Maritime Areas of Rocks Which Cannot Sustain Human Habitation or Economic Life of Their Own," *Netherlands Yearbook of International Law* vol. 21 (1990) pp. 139–181. A notorious case in point is the uninhabited island of Rockall, 300 miles off the coast of Scotland, where the UK eventually had to give up its claim of a 200-mile EEZ under UNCLOS; *cf.* I. Brownlie, *Principles of Public International Law* (3rd edn. Oxford: Clarendon 1979) p. 230; and D. Stummel, "Rockall," in R. Bernhardt (ed.), *Encyclopedia of Public International Law* vol. 4 (Amsterdam: Elsevier 2000) pp. 265–266.
59. Lady Hazel Fox (notes 21 and 54 above), Lord Asquith (pp. **16–17**), and Lord Gore–Booth (note 9).
60. Statement by the Parliamentary Undersecretary of State, FCO, *House of Commons Debates* vol. 423 (July 7, 2004) col. 292WH, *British Yearbook of International Law* vol. 75 (2004) p. 669, citing the FCO–commissioned study by Posford Haskoning Consultants, *Feasibility Study for the Resettlement of the Chagos Archipelago: Phase 2B* (Peterborough: Royal Haskoning 2002); also summarized in *British Yearbook of International Law* vol. 75 (2004) pp. 663–664, 668 and 672, and vol. 77 (2006) p. 638. The conclusions of the study were, however, challenged in a subsequent report by J. Howell (ed.), *Returning Home: A Proposal for the Resettlement of the Chagos Islands* (London: UK Chagos Support Association, March 2008), which in turn has been critically reviewed by J.R. Turner *et al.*, *An Evaluation of 'Returning Home—A Proposal for the Resettlement of the Chagos Islands' (Howell Report)* (London: Chagos Conservation Trust, June 2008). See also Chapter 5 (note 77) and Chapter 6 (note 11).
61. Letter dated January 18, 2006, written at the request of UK government representatives for use in the high court proceedings; quoted by Lord Justice Hooper (note 62, para. 97) as highlighting the risk of "terrorists infiltrating the island and by the use of missiles and electronic devices compromising the security of Diego Garcia." See also Chapter 6 (note 9).
62. *The Queen (ex parte Bancoult) vs. Foreign and Commonwealth Office*, [2006] *England and Wales High Court Reports* (Admin.) 1038; see S. Farran, "Prerogative Rights, Human Rights and Island People: The Pitcairn and Chagos Island Cases," *Public Law* vol. 50 (January 2007) pp. 414–424.
63. *Secretary of State for Foreign and Commonwealth Affairs vs. The Queen (on the application of Bancoult)*, [2007] *England and Wales Court of Appeal Reports* (Civ.) 498, [2007] 3 *Weekly Law Reports* 768; see S. Allen, "Looking Beyond the Bancoult Case: International Law and the Prospect of Resettling the Chagos Islands," *Human Rights Law Review* vol. 7 (2007) pp. 441–482; A.W. Bradley, *'The Widest Powers Appropriate to a Sovereign': Reflections*

NOTES—CHAPTER 2

on Removal of the Chagos Islanders to Make Way For the U.S. Base on Diego Garcia, (Oxford University: Institute of European and Comparative Law, June 7, 2007; and University of Connecticut School of Law, October 3, 2007; on-line <http://www.uconn.edu/faculty/workshops/public/abradley-20071003.pdf>); and R. Moules, "Judicial Review of Prerogative Orders in Council: Recognising the Constitutional Reality of Executive Legislation," *Cambridge Law Journal* vol. 67 (2008) pp. 12–15.

64. Leave to appeal granted on October 29, 2007; see the press reports by M. Fletcher, *Times* (London, online November 9, 2007), and D. Campbell, "Law Lords Hold Key to Islanders' Dream of Return to Paradise," *Guardian* (London, online May 12, 2008).
65. See D. Snoxell, "Expulsion from Chagos: Regaining Paradise," *Journal of Imperial and Commonwealth History* vol. 36 (2008) pp. 119–129); D. Campbell, "Diego Garcia: Chagos Islands Return 'Puts US Base at Risk'," *Guardian* (London, online July 1, 2008); S. Carey, "Chagos and the Law Lords," *New Statesman* (London, online July 1, 2008), reprinted in *Mauritius Times* (Port Louis, online 4 July 2008); and D. Vine, "Decolonizing Britain in the 21st Century," *Anthropology Today* vol. 24 No. 4 (August 2008) pp. 26–28.
66. *Foreign and Commonwealth Office: Managing Risks in the Overseas Territories*, 7th Report of the House of Commons Committee of Public Accounts, Session 2007–08, Report No. HC 176 (London: HM Stationery Office 2008) p. 22. During oral hearings on December 10, 2007 (with the participation of the BIOT Commissioner), Austin Mitchell MP suggested "perhaps to submit a bill to the Americans" (p. 23).
67. See notes 38–39. More than 1,000 pages of documents from the proceedings are available on–line at the website of the plaintiffs' attorney, American University Professor Michael E. Tigar, <http://homepage.ntlworld.com/jksonc/5_DiegoGarcia.html>.
68. An application to this effect had already been introduced by the legal representatives of the Islanders (*Sheridan Solicitors*) in April 2005, after the Court of Appeals decision (note 49), but was adjourned pending further UK domestic proceedings; the application was reactivated after the Law Lords' decision, in January 2009. On procedure see K. Reid, *A Practitioner's Guide to the European Convention on Human Rights* (3rd edn. London: Sweet & Maxwell 2008); T. Barkhuysen and M. van Emmerik, "Legitimacy of European Court of Human Rights Judgments: Procedural Aspects," in N. Nuls et al. (eds.), *The Legitimacy of Highest Courts' Rulings: Judicial Deliberations and Beyond* (Cambridge: Cambridge University Press 2009) pp. 437–450.
69. Moor and Simpson (Chapter 1, note 23) pp. 188–189, 193; and P.H. Sand, "Diego Garcia: British-American Legal Blackhole in the Indian Ocean?," *Journal of Environmental Law* vol. 21 (2009) pp. 113–139; cf. R. Gifford, "The Chagos Islands: The Land Where Human Rights Hardly Ever Happen," *Law, Social Justice and Global Development Journal* vol. 7 (2004 online). On similar "legal black hole" theories concerning the U.S. base at Guantánamo Bay, see Chapter 1 (note 67); see also pp. **59**, **64**, and Chapter 3 (note 36), Chapter 4 (note 45).
70. See Chapter 1 (note 60).
71. See Chapter 1 (note 63).
72. See Chapter 1 (note 65).
73. See note 18 above, and Chapter 1 (notes 62 and 64). "Derogations" to prevent or restrict application of the European Convention on Human Rights (note 18) in UK dependent territories were drafted by a coordinating committee in the Foreign and Commonwealth Office; see Simpson (note 18) p. 876, and E. Bates, "Avoiding Legal Obligations Created by Human Rights Treaties," *International and Comparative Law Quarterly* vol. 57 (2008) pp. 751–788, at p. 753. On UK practice since 1967 (following a traditional 'dualist' doctrine with regard to the transformation of treaty law into domestic law), see A.I. Aust, *Modern Treaty Law and Practice* (2nd edn. Cambridge: Cambridge University Press 2007) p. 206, and Lord Hoffmann's opinion in the *Bancoult 2* case, para. 66 (annex X, p. **143**). The question of

conformity with international treaty law was deliberately "left open" in the dissenting opinion of Lord Mance (paras. 142 and 145); but *cf.* critically Allen (Chapter 1, note 64) at p. 689.

74. The European Convention on Human Rights (adopted at Strasbourg on November 4, 1950, 213 United Nations Treaty Series 222, in force for the United Kingdom since September 3, 1953) is now considered part of British domestic law, pursuant to the Human Rights Act of 9 November 1998 (Act to give further effect to rights and freedoms guaranteed under the European Convention on Human Rights), in force October 2, 2000 (1998 Statutes, Ch. 42); see S. Besson, "The Reception Process in Ireland and the United Kingdom," in H. Keller and A.S. Sweet (eds.), *A Europe of Rights: The Impact of the ECHR on National Legal Systems* (Oxford: Oxford University Press 2008) pp. 31–106. According to UK declarations registered with the Council of Europe as of February 23 , 2006, the geographical scope of application of that convention has been extended to almost all British overseas territories *except* BIOT and the Pitcairn Islands; see p. **64**, and Moor and Simpson (Chapter 1, note 23) pp. 154 and 156, Simpson (note 18) pp. 1097–1098. It has been pointed out, though, that the United Kingdom had originally (by notification in 1953, under art. 56 of the convention) already extended the treaty to the territory of Mauritius, which at that time still included Diego Garcia; see the submission dated October 12, 2007 by R. Gifford to the House of Commons Select Committee on Foreign Affairs, *Overseas Territories: Seventh Report of Session 2007–08* (Chapter 1, note 64) pp. Ev105–11, at p. 111. Moreover, the Human Rights Act has been held applicable outside UK territory, regardless of geographical extensions, provided the UK has sufficient control to enable it to secure Convention rights to everyone; see The Queen (Al–Skeini and others) vs. Secretary of State for Defence, [2007] *United Kingdom House of Lords Decisions* 26, [2007] 3 *Weekly Law Reports* 33; and D. Feldman, "The Territorial Scope of the Human Rights Act 1998," *Cambridge Law Journal* vol. 67 (2008) pp. 8–10.

75. *House of Commons Debates* vol. 423 (July 7, 2004), col. 289WH, reprinted in *British Yearbook of International Law* vol. 75 (2004) p. 665; *cf.* note 21 above, but see also Chapter 1 (note 64) on the debate in the UN Human Rights Committee.

76. D.R. Snoxell, letter to the editor, *Times* No. 69022 (London, May 26, 2007) p. 22. See also Lord John Hatch of Lusby, *House of Lords Debates* vol. 436 (November 11, 1982), col. 401 ("a most shameful incident in British history"); Simpson (note 18) p. 1100 ("a particularly scandalous case of abuse of human rights"); Palmer (note 47) p. 236 ("a sorry Chapter in colonial history"); D. Vine, "The Impoverishment of Displacement: Models for Documenting Human Rights Abuses and the People of Diego Garcia," *Human Rights Brief* vol. 13 (2006) pp. 21–32; A. Perreau-Saussine, "British Acts of State in English Courts," *British Yearbook of International Law* vol. 78 (2007) pp. 176–254, at p. 246 ("highlights the injustice of the British state's actions towards the Chagos islanders"); and S. Carey, "Britain's Worst Gift," *New Statesman* (London, online June 23, 2008): "Without doubt this episode must rank as one of the most shameful and squalid in recent colonial history."

77. Statements during the 1975 congressional hearings (notes 26–27 above); also quoted by Winchester (Chapter 1, note 16) p. 224.

3 Power Politics: "Our" Ocean

1. Testimony of Ronald I. Spiers, director of the U.S. State Department's Bureau of Politico-Military Affairs, in *The Indian Ocean: Political and Strategic Future*: Hearings before the Subcommittee on National Security Policy and Scientific Developments, U.S. House of Representatives Committee on Foreign Affairs, 92nd Congress, 1st Session (July 1971)

p. 165; see also R.I. Spiers, "U.S. National Security Policy and the Indian Ocean Area," *Department of State Bulletin* vol. 65 No. 1678 (1971) pp. 199–203. It was R. Spiers who five years later, as U.S. *chargé d'affaires* in London in July 1976, signed the UK-U.S. Agreement converting Diego Garcia into a major naval support base (appendix IV, p. **91**).
2. See D. Widome, "The List: The Six Most Important U.S. Military Bases," *Foreign Policy* No. 21 (May 13, 2006); and U.S. Department of Defense, *Base Structure Report: Fiscal Year 2007 Baseline* (Washington, D.C.: DoD 2007) p. 78: total plant replacement value (PRV) in Diego Garcia given as $2.541 billion. See also R.E. Harkavy, *Bases Abroad: The Global Foreign Military Presence* (Oxford: Oxford University Press 1989) pp. 49, 310; S. Winchester, "La plus grande base américaine du monde: Diego Garcia, ses plages et ses super–bombardiers," *Courrier International* No. 573 (October 25–30, 2001) p. 53; and C.A. Lutz, "Bases, Empire, and Global Response," in C.A. Lutz (ed.), *The Bases of Empire: The Global Struggle Against US Military Posts* (London: Pluto Press 2008), pp. 1–44. Unlike a number of other locations, Diego Garcia has *not* been targeted for closure in the context of the "Global Posture Review" (GPR) of U.S. military bases abroad, presented to Congress by Defense Secretary Donald H. Rumsfeld in 2004; see Calder (Chapter 1, note 58); M.E. O'Hanlon, *Unfinished Business: U.S. Overseas Military Presence in the 21st Century* (Washington,D.C.: Center for a New American Security, June 2008); and C. Johnson, *Nemesis: The Last Days of the American Republic* (New York: Holt 2008) at p. 147. But see p. **59**.
3. See S.K.M. Panikhar, *India and the Indian Ocean: An Essay on the Influence of Sea Power on Indian History* (London: Allen & Unwin 1951); R. Hall, *Empires of the Monsoon: A History of the Indian Ocean and Its Invaders* (London: Harper Collins 1996); S. Bose, *A Hundred Horizons: The Indian Ocean in the Age of Global Empire* (Cambridge/MA: Harvard University Press 2006); and E. Chew, Crouching Tiger, Hidden Dragon: The Indian Ocean and the Maritime Balance of Power in Historical Perspective, Working Paper No. 144 (Singapore: Rajaratnam School of International Studies 2007).
4. *Mare Nostrum* ["our sea"] was the term applied by the Romans (and later, by the Italian fascists) to the Mediterranean Sea. In a different contemporary perspective, the term has also been used to postulate global access to the world's oceans by all human beings; see P. Allott, "*Mare Nostrum*: A New International Law of the Sea," in J.M. Van Dyke *et al.* (eds.), *Freedom for the Seas in the 21st Century: Ocean Governance and Environmental Harmony* (Washington,D.C.: Island Press 1993) pp. 49–71, at p. 56; and P. Allott, *Towards the International Rule of Law: Essays in Integrated Constitutional Theory* (London: Cameron May 2005) at p. 355.
5. See (Cdr.) D.W. Urish, "To Build a Link: The Seabees at Diego Garcia," *U.S. Naval Institute Proceedings* vol. 99 No. 4 (1973) pp. 101–104; and K. Harrison, "Diego Garcia: The Seabees at Work," *U.S. Naval Institute Proceedings* vol. 105 No. 8 (1979) pp. 53–61.
6. Bandjunis (Chapter 1, note 6) p. 55.
7. Bandjunis *ibid.* p. 216; L.W. Bowman and J.A. Lefebvre, "The Indian Ocean: U.S. Military and Strategic Perspectives," in Dowdy and Trood (Chapter 1, note 58) pp. 413–435, at p. 423. *Penta-Ocean* was subsequently involved in a lawsuit brought by the U.S. government against 26 construction companies for alleged bid-rigging at the U.S. Navy's Atsugi base near Tokyo in 2002.
8. On a "cost-plus-award-fee" basis (CPAF, i.e. cost-reimbursable); as to the contractors, see the background note by J.M. Carter, "The Merchants of Blood: War Profiteering from Vietnam to Iraq," *CounterPunch* (online December 11, 2003); and P. Chatterjee, *Halliburton's Army: The Long, Strange Tale of a Private, Profitable, and Out-of-control Texas Oil Company* (New York: Nation Books 2009). *Brown & Root* was bought by *Halliburton* in 1962; under the chairmanship (1980–1992) of British engineer Sir Richard Morris CBE, it merged with M.W. Kellogg to become *Kellogg Brown & Root* (*KBR* since 1997); see Sir Richard's obituary, *Times* No. *69384* (London, online July 24, 2008) p. 61.
9. Awarded on July 14, 1981; see Edis (Chapter 1, note 3) p. 91, and Bandjunis (Chapter 1, note 6) pp. 219–220.

10. Completed in January 1987; see T. Tucker [Capt., Navy Civil Engineer Corps] and B.T. Doughty [RBRM], "Naval Facilities, Diego Garcia, British Indian Ocean Territory: Management and Administration," *Proceedings of the Institution of Civil Engineers* (Maritime Engineering Group: Part 1) vol. 84 (1988) pp. 191–215, at p. 209 (stating that "security considerations preclude providing a full description of the works").

11. U.S. General Accounting Office, *Further Improvements Needed in Navy's Oversight and Management of Contracting For Facilities Construction On Diego Garcia*, Report to the Secretary of Defense GAO/NSIAD-84-62 (Washington,D.C., May 23, 1984); Bandjunis (Chapter 1, note 6) p. 220.

12. Firm-fixed-price, indefinite-quantity fee provisions contract, operated since 1999 by *DG21 LLC*, a joint venture between *Lockheed Martin* of Bethesda/MD [the world's largest defence contractor] and *Day & Zimmermann* Munition & Defense Inc. of Philadelphia/PA, together with *First Support Services Inc.* of Dallas/TX, and with a 24.5 per cent share by UK engineering consultants *W.S. Atkins PLC* of Epsom/Surrey [one of whose directors is retired Admiral Michael C. Boyce, Baron Boyce of Pimlico GCB OBE DL, former First Sea Lord of the Royal Navy]. The cumulative contract (No. N62742-06-D-4501, renewed on July 30, 2008) currently stands at $467.7 million; see *Defense Industry Daily* (online August 4, 2008).

13. At a daily rate of $12,550, under DoD contract No. N00033–05–C–5500 (renewed in September 2008); on the return voyage, wastes have been shipped out from the base, initially for disposal in the Philippines (see Tucker and Doughty, note 10, p. 214) and to the Defense Reutilization Marketing Office in Fort Lewis/WA. In 2006, an accumulated 10 million pounds (4,500 tonnes) of scrap metal on Diego Garcia were put up for commercial sale (netting $133,000) and have since been shipped to the United States by *Big Iron Trading* Co.; see *Navy Newsstand* Release No. NNS061013–09 (October 13, 2006).

14. The corresponding minimum share for U.S. firms is 60 per cent; see appendix VII (p. **109**), and *cf.* for example, the arrangements under the *DG21 LLC* joint venture (note 12). Corresponding tax revenues for the UK Treasury from British contractors may safely be estimated at a total of more than $20 million since 1987.

15. See Bowman and Lefebvre (note 7) at p. 584/n. 28; Patel (Chapter 2, note 33) p. 115; Vine (Chapter 1, note 6) p. 189; and Jeremy Corbin MP, statement during parliamentary debate of the British Overseas Territories Bill (Chapter 2, note 44), *House of Commons Debates* vol. 375 (November 22, 2001) col. 516 ("Îlois barred from civilian employment"). Over the past 25 years, only 3 native Chagossians found employment at the U.S. base.

16. Bandjunis (Chapter 1, note 6) p. 256 table 11 (showing U.S. Diego Garcia personnel strength on September 30, 1984, as 5,943). In 2001, there were about 3,500 Americans, 1,250 Filipinos, 125 Mauritians and 30 British nationals in residence; see "Diego Garcia Under Occupation," in: *Diego Garcia in Times of Globalisation* (Port Louis: LALIT 2001) pp. 155–159.

17. C. Sheppard, "The Coral Fauna of Diego Garcia Lagoon, Following Harbour Construction," *Marine Pollution Bulletin* vol. 11 (1980) pp. 227–230, at p. 228.

18. See C.A. Johnson, *The Sorrows of Empire: Militarism, Secrecy, and the End of the Republic* (New York: Metropolitan Holt 2004) p. 221. On December 16, 1985, *USS Saratoga* [CV–60] was the first aircraft carrier to dock at the new "Alpha" wharf; in June 1987, *USS Constellation* [CV–64, the world's largest conventionally powered warship] conducted flight operations while anchored in Diego Garcia; followed in December 1987 by *USS Long Beach* [CGN–9, the first nuclear–powered guided missile cruiser]. See T.A. Morris, "The True History of Diego Garcia," <http://www.zianet.com/tedmorris/dg/realhistory.html> (last accessed December 26, 2008); and Edis (Chapter 1, note 3) p. 92. On the most recent wharf expansions for use by nuclear–powered *SSGN* attack submarines see note 47.

19. See Oraison (Chapter 1, note 18) p. 140; and Erickson *et al.* (Chapter 1, note 11) p. 16. It replaced the smaller runway previously built from coral cement by the "Seabees"; see B.T. Doughty, discussion in *Proceedings of the Institution of Civil Engineers* vol. 86 (1989) p. 417.

NOTES—CHAPTER 3

20. (Cdr.) J. Clementson, "Diego Garcia," *Journal of the Royal United Services Institute for Defence and Security Studies (RUSI)* vol. 126 No. 6 (1981) p. 33, at pp. 36–38; Bowman and Lefebvre (note 8) p. 423; and R. O'Rourke, *Navy–Marine Corps Amphibious and Maritime Prepositioning Ship Programs: Background and Oversight Issues*, CRS Report for Congress RL32513 (Washington,D.C.: Congressional Research Service November 15, 2004, updated 10 July 2007) p. 4.
21. Winchester (Chapter 1, note 12) p. 51, reprinted in Mancham (Chapter 1, note 11) p. 45; see also Winchester (note 2), and Bandjunis (Chapter 1, note 6) pp. 256–257.
22. See J. Khan, "Diego Garcia: The Militarization of an Indian Ocean Island," in R. Cohen (ed.), *African Islands and Enclaves* (Beverly Hills/CA: Sage 1983) pp. 169–174; Rais (Chapter 1, note 35) at pp. 80–87; and J. Houbert, "The West in the Geopolitics of the Indian Ocean and India," in D. Rumley and S. Chaturvedi (eds.), *Geopolitical Orientations, Regionalism and Security in the Indian Ocean* (New Delhi: South Asian Publishers 2004).
23. See Bezboruah (Chapter 1, note 6) pp. 109–118, citing senators Mike Mansfield [until his departure from the Senate as U.S. ambassador to Japan in 1977], Claiborne Pell, Edward Kennedy and John Culver, in particular. Along similar lines, see the warnings expressed by the *Stockholm International Peace Research Institute* (SIPRI), in its *1975 Yearbook* (Stockholm: Almquist & Wiksell 1976) pp. 81–82. See also Calder (Chapter 1, note 58) p. 185; and R.D. Johnson, *Congress and the Cold War* (Cambridge: Cambridge University Press 2006) pp. 212–213.
24. E.g., see the statements by Admiral Elmo R. Zumwalt on April 11, 1974, *Briefings on Diego Garcia and Patrol Frigate*, Hearings before the U.S. Senate Committee on Foreign Relations, 93rd Congress, 2nd Session (Washington,D.C.: Government Printing Office 1974) p. 12, referring to Soviet plans for a naval base in Somalia. See also A.J. Cottrell and R.M. Burrell, "The Soviet Navy and the Indian Ocean," *Strategic Review* vol. 2 No. 3 (1974) pp. 25–35; (Adm.) T.H. Moorer and A.J. Cottrell, "The Search for U.S. Bases in the Indian Ocean: A Last Chance," *Strategic Review* vol. 8 No. 1 (1980) pp. 30–31; Cottrell and Associates (Chapter 1, note 58) p. 125; and Jawatkar (Chapter 1, note 28) p. 138 ("bogey of Soviet threat").
25. Testimony of Rear Admiral Charles D. Grojean, director of the U.S. Navy's Politico–Military Policy Division, before the House of Representatives' Armed Services Subcommittee on Military Construction (on June 4, 1974), as quoted in Bandjunis (Chapter 1, note 6) p. 95. See also the statements by State Department official Seymour Weiss (Chapter 2, note 27), reprinted as "US Interests and Activities in the Indian Ocean," *Department of State Bulletin* vol. 70 No. 1815 (1974) pp. 371–379, at p. 375 ("this does not represent the development of a major new base"); and by Lord Skelmersdale [Roger Bootle–Wilbraham, 7th Baron Skelmersdale] on behalf of the UK Government, *House of Lords Debates* vol. 436 (November 11, 1982), col. 408 (support facility, "not a military base"). But see note 2.
26. On the Australian position—until the change of government in 1976—see Jawatkar (Chapter 1, note 28) pp. 192–194; Bezboruah (Chapter 1, note 6) pp. 200–204; and H.S. Albinski, "Australia, New Zealand, and Indian Ocean Security," in Dowdy and Trood (Chapter 1, note 58) pp. 356–377, at p. 367. See also P. Hayes et al., *American Lake: Nuclear Peril in the Pacific* (Victoria/Australia: Penguin Books 1986); and the consultations on Diego Garcia between U.S. Secretary of State Henry Kissinger and UK Cabinet Secretary Sir John Hunt, recorded in a "Memorandum of Conversation" at the White House on April 26, 1974; *Records of Henry Kissinger 1973–1977*, U.S. National Archives Record Group 59, Box 7, Apr 1974 Nodis Memcons (marked "secret," declassified on August 1, 2000).
27. See S.S. Harrison, "India, the United States, and Superpower Rivalry in the Indian Ocean," in Harrison and Subrahmanyam (Chapter 1, note 24) pp. 246–286, at pp. 258–259; and generally W. Arkin and R. Fieldhouse, *Nuclear Battlefields: Global Links in the Arms Race* (Cambridge/MA: Ballinger 1985).

28. On the Blackpool motion, and the lingering controversy over the U.S. duty to "consult" Britain over any deployment of nuclear weapons at Diego Garcia, see J. Larus, "Diego Garcia: Political Clouds Over a Vital U.S. Base," *Strategic Review* vol. 10 No. 3 (1982) pp. 44–55, and Larus (Chapter 1, note 58) pp. 444–448.
29. Quoted in J.P. Anand, "Diego Garcia Base," *Institute for Defence Studies and Analyses (IDSA) Journal* vol. 11 No. 3 (1979) pp. 58–85, at p. 71; see also N.D. Palmer, "South Asia and the Indian Ocean," in Cottrell and Burrell (Chapter 1, note 58) pp. 235–250, at p. 245; Jawatkar (Chapter 1, note 28) pp. 174–191; and W.K. Andersen, "Emerging Security Issues in the Indian Ocean: An American Perspective," in Harrison and Subrahmanyam (Chapter 1, note 24) pp. 12–83, at p. 23. However, Erickson *et al.* (Chapter 1, note 11) p. 21, suggest that India has now "lost its aversion" to the base in the face of a perceived Chinese threat, as evidenced by its participation in joint naval exercises at Diego Garcia in 2001/2004.
30. UN General Assembly Resolution 2832 (XXVI) of December 16, 1971, adopted by 61 votes in favour and none opposed, with 55 abstentions. See D. Kaushik, *The Indian Ocean: Towards a Peace Zone* (New Delhi: Vikas Publications 1972); B. Buzan, "Naval Power, the Law of the Sea, and the Indian Ocean as a Zone of Peace," *Marine Policy* vol. 5 (1981) pp. 194–204; D. Braun, *The Indian Ocean: Region of Conflict or Zone of Peace?* (London: Hurst 1983); M.C.W. Pinto, "The Indian Ocean as a Zone of Peace," in R.W. Byers (ed.), *The Denuclearization of the Oceans* (New York: Palgrave Macmillan 1986) pp. 145–156; K. Subrahmanyam, "Arms Limitation in the Indian Ocean: Retrospect and Prospect," in Harrison and Subrahmanyam (Chapter 1, note 24), pp. 223–245; and P. MacAlister-Smith, "Zones of Peace," in R. Bernhardt (ed.), *Encyclopedia of Public International Law* vol. 4 (Amsterdam: Elsevier 2000) pp. 1621–1633, at p. 1625.
31. UN General Assembly Resolution 2992 (XXVII) of December 15, 1972. See V.S. Deshpande, "Indian Ocean as a Peace Zone: Evolving the Legal Process," *Indian Journal of International Law* vol. 14 (1974) pp. 160–168; Jawatkar (Chapter 1, note 28) pp. 215–225; and P. Towle, "The United Nations Ad Hoc Committee on the Indian Ocean: Blind Alley or Zone of Peace?," in L.W. Bowman and I. Clark (eds.), *The Indian Ocean in Global Politics* (Boulder/CO: Westview 1981) pp. 207–221.
32. UN Documents A/45/213, A/45/214, and A/45/215, of April 11, 1990.
33. MacAlister-Smith (note 30) p. 1626.
34. African Nuclear-Weapon-Free Zone Treaty, *International Legal Materials* vol. 35 (1996) p. 698; endorsed by UN General Assembly Resolution 50/78 of November 6, 1995.
35. See N. Pelzer, "Nuclear-Free Zones," in R. Bernhardt (ed.), *Encyclopedia of Public International Law* vol. 3 (Amsterdam: Elsevier 1997) pp. 705–714, at p. 712; and Ollivry (Chapter 1, note 39) at p. 93. See also the explanatory memorandum on the Pelindaba Treaty submitted to the UK Parliament by the FCO in December 2000 (Command Paper 3498), para. iv/a/ii.
36. As quoted by J. Goldblat, "Nuclear–Weapon–Free Zones: A History and Assessment," *Nonproliferation Review* vol. 4 No. 3 (1997) pp. 18–32, at p. 26; and S.J. Forsberg, Island at the Edge of Everywhere: A History of Diego Garcia (Huntsville/TX: Sam Houston State University thesis 2005) p. 53.—The "inapplicable treaty" doctrine parallels the "legal black–hole" doctrines referred to in Chapter 1 (note 67).
37. Antarctic Treaty, adopted at Washington/D.C. on December 1, 1959 (in force June 23, 1961), 402 United Nations Treaty Series 71, Article IV(1): "Nothing contained in the present Treaty shall be interpreted as...prejudicing the position of any Contracting Party as regards its recognition or non-recognition of any other State's right of or claim to territorial sovereignty in Antarctica." E.g., a sector of the *British Antarctic Territory* (between 25°W and 74°W) is also claimed by Argentina. See also p. **66**.
38. See D. Fischer, "The Pelindaba Treaty: Africa Joins the Nuclear Free World," *Arms Control Today* vol. 25 No. 10 (1995) pp. 9–20, at p. 10 (quoting an official in the US Arms Control and Disarmament Agency); and M.E. Rosen [Capt., international law attorney in the US Navy Judge Advocate General's Corps], "Nuclear-Weapon-Free Zones: Time for a Fresh

Look," *Duke Journal of Comparative and International Law* vol. 8 (1997) pp. 29–78, at p. 48. But see, on the reasons for U.S. failure to ratify, K.K. Schonberg, "The Generals' Diplomacy: U.S. Military Influence in the Treaty Process, 1992–2000," *Seton Hall Journal of Diplomacy and International Relations* vol. 3 (2002) pp. 68–83, at p. 80.
39. See Zumwalt, note 24.
40. See Erickson et al. (Chapter 1, note 11) p. 17, and Chapter 5 (notes 71–72). *Cf.* D.F. Doyon, "Middle East Bases: Model for the Future," in J. Gerson and B. Birchard (eds.), *The Sun Never Sets: Confronting the Network of Foreign U.S. Military Bases* (Boston: South End 1991) pp. 257–307, at p. 294, noting that one U.S. congressional aide facetiously described Diego Garcia—off the record—as so heavily militarized that "we're waiting for it to sink."
41. E.g., see Andersen (note 29) p. 23; Anand (note 29) p. 72; R.W. Jones, "Ballistic Missile Submarines and Arms Control in the Indian Ocean," *Asian Survey* vol. 20 No. 3 (March 1980); Larus (Chapter 1, note 58) pp. 435 and 444; and Bhatt (Chapter 1, note 11) p. 39; see also the political debate in Britain, note 28. Verification of any nuclear weapons deployments is notoriously difficult; the "Brookings Audit" (S.I. Schwartz ed., *Bases and Facilities with Significant Current or Historical U.S. Nuclear Weapons or Naval Nuclear Propulsion Missions*, Washington/D.C.: Brookings Institution, updated as of August 2002) shows *no* figures for the Diego Garcia base.
42. See appendix IV (p. 92), Article 3.
43. P. Beaumont and C. Urquhart, "Israel Deploys Nuclear Arms in Submarines," *Observer* (London, October 12, 2003), quoting Israeli and U.S. officials interviewed by the *Los Angeles Times*. According to Article 4(c) of the 1976 agreement (appendix IV, p. 93), the operation would have required the prior consent of the UK government.
44. EU Doc. 8675/2/98/Rev.2, adopted on June 8, 1998; see generally K. Holm, "Europeanising Export Controls: The Impact of the European Union Code of Conduct on Arms Exports in Belgium, German and Italy," *European Security* vol. 15 (2006) pp. 213–234. The related "Common Military List of the European Union: Equipment Covered by the European Union Code of Conduct on Arms Exports," adopted on March 19, 2007, explicitly lists "vessels of war (surface and underwater)"; Official Journal of the European Union [2007] L88/58 (ML9).
45. Treaty on the Non-Proliferation of Nuclear Weapons, adopted on July 1, 1968 (in force March 5, 1970), 729 United Nations Treaty Series 161. See H. Müller, D. Fischer and W. Kötter, *Nuclear Non-Proliferation and Global Order* (Oxford: Oxford University Press 1994); K. Ioannou, "Non-Proliferation Treaty," in R. Bernhardt (ed.), *Encyclopedia of Public International Law* vol. 3 (Amsterdam: Elsevier 1997) pp. 622–627; and M.I. Shaker, "The Evolving International Regime of Nuclear Non-Proliferation," *Hague Academy of International Law: Collected Courses* vol. 321 (2006) pp. 9–202.
46. See K. Bailey and R. Rudney (eds.), *Proliferation and Export Controls* (Lanham/MD: University Press of America 1993); C. Ahlström, "Non-Proliferation of Ballistic Missiles: The 2002 Code of Conduct," *Stockholm International Peace Research Institute Yearbook* (SIPRI 2003) pp. 749–759; and W. Boese, "Israel Allegedly Fielding Sea-Based Nuclear Missiles," *Arms Control Today* vol. 33 No. 9 (2003) p. 26.
47. Wharf construction works to be completed in April 2009; contract No. N62742-07-C-1313, *Defense Industry Daily* (online April 4, 2007). The 23,000-ton tender *USS Emory S. Land* (AS-39), which will serve as a floating shipyard to repair and supply submarines and surface vessels at Diego Garcia from 2009 onwards, had to leave its former base at La Maddalena/Italy in September 2007 because of local political protests; see Erickson et al. (Chapter 1, note 11) pp. 18–20.
48. Treaty on the Reduction and Limitation of Strategic Offensive Arms (START), with Protocol of Inspection and Continuous Monitoring Activities Relating to the START Treaty, bilateral U.S.–Russian agreement signed at Moscow on July 31, 1991 (due to expire on December 5, 2009); 31 *International Legal Materials* 246. See D.B. Thomson, *The*

Strategic Arms Reduction Treaty and its Verification (Los Alamos/NM: National Lab 1992); and T. Bruha, "Strategic Offensive Arms, Treaties on Reduction and Limitation (START)," in R. Bernhardt (ed.), *Encyclopedia of Public International Law* vol. 4 (Amsterdam: Elsevier 2000) pp. 705–710.

49. A. Diakov and E. Miasnikov, "RESTART: The Need for a New U.S. Russian Strategic Arms Agreement," *Arms Control Today* vol. 36 No. 7 (2006) pp. 6–11, at p. 9, citing A. Scutro, "Balance of Sub Fleet to Swing Toward the Pacific," *Navy Times* (February 20, 2006). Hence, Diego Garcia will also qualify as a disarmament 'black hole'.

50. As announced by Pentagon spokesman B. Whitman; see *CNN-Online* (July 7, 2008). The uranium "yellowcake" material originated from the remnants of Iraq's former nuclear facility in Tuwaitha (bombed out of operation by Israel in 1981), and was sold by the Government of Iraq to *Cameco Corporation* in Ontario/Canada—the world's largest producers of radioactive material—for enrichment processing and use in nuclear power plants; see also Erickson *et al.* (Chapter 1, note 11) p. 17 footnote 79.

51. See Chapter 5 (notes 40 and 59).

52. On the problems of the Guam-based submarine *USS Houston* (SSN-713), see *International Herald Tribune: Asia-Pacific*/Associated Press (online, August 30, 2008); *cf.* M. Yamaguchi, "Nuclear Sub's Radiation Leak Revealed," *Seattle Times* (online, August 8, 2008).

53. E.g., the published results of the 2006 scientific expedition under the auspices of the BIOT Administration, which did take water quality samples from parts of the Diego Garcia lagoon, did not include radiation measurements; see Chapter 5 (note 55).

54. Testimony of Admiral Thomas H. Moorer, chairman of the U.S. joint chiefs of staff, before the Senate's Armed Services Committee (on March 12, 1974), confirming earlier statements by State Department representative Seymour Weiss (Chapter 2, note 27); as quoted in Bandjunis (Chapter 1, note 6) p. 80.

55. As recorded by Morris, note 18.

56. Newsom-Wilkinson letter of June 21, 2000 (Chapter 2, note 46) p. 2; and Edis (Chapter 1, note 3) p. 96. Diego Garcia had also served as a support base—with or without British consent—for the ill-starred attempt at rescuing the U.S. hostages in Iran in April 1980 (*Operation Eagle Claw*); see J.L. Thigpen, *The Praetorian Starship: The Untold Story of the Combat Talon* (Maxwell/AL: Air University Press 2001) p. 212; legal analysis in T.L. Stein, "Contempt, Crisis, and the Court: The World Court and the Hostage Rescue Attempt," *American Journal of International Law* vol. 76 (1982) pp. 499–531.

57. W.M. Arkin, "Calculated Ambiguity: Nuclear Weapons and the Gulf War," *Washington Quarterly* vol. 19 No. 4 (1996) pp. 3–18, at p. 10 (quoting former U.S. Secretary of State James Baker); and Rosen (note 38) p. 48.

58. Edis (Chapter 1, note 3) pp. 95–97; Calder (Chapter 1, note 58) p. 186; D.L. Haulman, "Footholds for the Fighting Force," *Air Force Magazine* vol. 89 No. 2 (2006) pp. 76–79; and Morris (note 18), noting that on April 20, 2007, three U.S. Air Force colonels received the *Air Force Productivity Excellence Award* for their discovery.

59. I. Bruce, "Secret Move to Upgrade Air Base for Iran Attack Plans," *Herald* (Glasgow, October 29, 2007); Agence France Press, "Experts: Pentagon Pushing Bunker Buster for Iran Use," *Defense News* (October 25, 2007); and Erickson *et al.* (Chapter 1, note 11) p. 17 footnote 78. See also Chapter 5 (note 83).

60. E.g., when releasing its *Diego Garcia Integrated Natural Resources Management Plan 2005* (see Chapter 5, note 41), the U.S. Navy deletes narrative legend from most maps, and all "sensitive" sections from the text; for a notable exception see Map 2 (p. **12**). The most informative private Internet website available, by Diego Garcia veteran Ted A. Morris [Lt. Col. USAF ret., note 18], was temporarily "sanitized" after September 11, 2001, but has since been largely rebuilt.

61. E.g., a *Reuters* journalist's freedom-of-information request for the 2004 valedictory dispatch of the former UK high commissioner to Mauritius and deputy BIOT commissioner

(ambassador David R. Snoxell, notorious for his criticism of UK and U.S. policies on Diego Garcia) was declined by the Foreign and Commonwealth Office "on the grounds that it would be prejudicial to both Anglo-Mauritian and Anglo-American relations," and hence will not become publicly available until 2034; Snoxell interview in the British Diplomatic Oral History Programme (Cambridge University: Churchill Archives Centre, November 19, 2007) p. 41. See also note 11; Chapter 1 (note 37); Chapter 5 (notes 26–28, 33, 43–44, 48, 84), and Chapter 6 (note 20).

4 Military Secrecy: Public Access Denied

1. See Chapter 2 (note 46).
2. See Forsberg (Chapter 3, note 36) p. 46; see also Hayes et al. (Chapter 3, note 26) pp. 439–446.
3. B.D. Nagchoudhary, preface to Jawatkar (Chapter 1, note 28) p. viii; on the strategic advantages of Diego Garcia's splendid isolation, see also Edis (Chapter 1, note 3) p. 98.
4. As quoted by MacAskill and Evans (Chapter 2, note 49) p. 3.
5. [Capt.] T. Tucker, discussion of the report referred to in Chapter 3 (note 10), *Proceedings of the Institution of Civil Engineers* vol. 86 (1989) p. 418. That statement brazenly disregards the 1975 congressional hearings and the related media criticism; see Chapter 2 (notes 26–29) and Chapter 3 (note 23).
6. Described as an "Anglo-American collection site" by R.J. Aldrich, "The Value of Residual Empire: Anglo-American Intelligence Co-operation in Asia after 1945," in R.J. Aldrich and M.F. Hopkins (eds.), *Intelligence, Defence and Diplomacy: British Policy in the Post-War World* (Ilford: Cass 1994) pp. 226–258, at p. 249. See also C. Andrew, "The Making of the Anglo-American Sigint Alliance," in H. Peake and S. Halperin (eds.), *In the Name of Intelligence: Essays in Honor of Walter Pforzheimer* (Washington, D.C.: NIBC Press 1994) pp. 105–118 and M.M. Aid, "The National Security Agency and the Cold War," in M.M. Aid and C. Wiebes (eds.), *Secrets of Signals Intelligence During the Cold War and Beyond* (London: Cass 2001) pp. 27–66.
7. There is a lot of conspiratorial literature on *Echelon*; among the more serious studies see J.T. Richelson and D. Ball, *The Ties That Bind: Intelligence Cooperation between the UKUSA Countries—the United Kingdom, the United States of America, Canada, Australia and New Zealand* (Boston: Allen & Unwin 1985) pp. 205–206 (Diego Garcia), and P.R. Keefe, *Chatter: Dispatches from the Secret World of Global Eavesdropping* (New York: Random House 2005) pp. 72–76, 115–121; *cf.* generally S. Chesterman, *Shared Secrets: Intelligence and Collective Security* (Sydney: Lowy Institute for International Policy 2006). An intelligence-collection role for Diego Garcia was first suggested in February 1970 by R.A. Frosch, Assistant Secretary of the Navy for Research and Development [later assistant executive director of the United Nations Environment Programme (UNEP) in Nairobi], mainly in the cold war context of anti-submarine strategy; see Bandjunis (Chapter 1, note 6) p. 41. *Cf.* R. Frosch, "Underwater Sound: Deep-Ocean Propagation," *Science* vol. 146 No. 3646 (1964) pp. 889–894; and Chapter 5 (note 78).
8. Originally established under a 1953 bilateral defense treaty with Ethiopia, operation of the large U.S. base in Kagnew was increasingly affected since the 1970s by the *Eritrean Liberation Front*'s guerrilla warfare (which eventually led to independence in 1991); Ethiopia terminated the Kagnew lease in 1977. See Aldrich (note 6 above) p. 248; Richelson and Ball (note 7 above) p. 204; Bowman and Lefebvre (chapter 3, note 7) p. 423; and J. Bamford, *Body of Secrets* (New York: Doubleday 2001) p. 160.
9. M. Rich, "NSA Diego Garcia: The Prelude," *Cryptolog: Newsletter of the Naval Cryptology Veterans Association* vol. 21 No. 1 (2000) p.1; Bamford (note 8 above) p. 165.

Notes—Chapter 4

10. J.T. Richelson, *The U.S. Intelligence Community* (Cambridge/MA: Ballinger 1985) pp. 200–201; Drower (Chapter 2, note 11) p. 12.
11. Richelson and Ball (note 7) p. 159; R.Z. George and R.D. Kline (eds.), *Intelligence and the National Security Strategist: Enduring Issues* (Washington D.C.: National Defense University 2005) p. 192. See also Harkavy (Chapter 3, note 2) p. 184; W.E. Burrows, *Deep Black: Space Espionage and National Security* (New York: Random House 1986) p. 201; E.C. Whitman, "SOSUS: The 'Secret Weapon' of Undersea Surveillance," *Undersea Warfare* vol. 7 No. 2 (2005); J.M. Di Mento, *Beyond the Water's Edge: United States National Security and the Ocean Environment* (Medford/MA: Fletcher School of Law and Diplomacy 2006) pp. 256, 532; and see generally the UK-U.S. Memorandum of Understanding Concerning Ocean Surveillance Information Systems, US Treaties and Other International Acts Series No. 12735 (November 5, 1985, as amended/extended on February 21/March 15, 1996).
12. Forsberg (Chapter 3, note 36) p. 49, noting that the project was declassified by the Navy in 1998. The Naval Security Group Department's facility was formally "disestablished" in September 2005, continuing with contractor support as "Naval Computer and Telecommunications Station, Far East Detachment Diego Garcia."
13. See F. Francioni, "Il caso del 'Pueblo' e le norme internazionali sullo spionaggio," *Diritto Internazionale* vol. 23 (1969) pp. 319–37; I.D. de Lupis, "Foreign Warships and Immunity for Espionage," *American Journal of International Law* vol. 78 (1984) pp. 53–75; A.P. Rubin, "Some Legal Implications of the Pueblo Incident," *International and Comparative Law Quarterly* vol. 18 (1996) pp. 961–970; J. Kokott, "Pueblo Incident," in R. Bernhardt (ed.), *Encyclopedia of Public International Law* vol. 3 (Amsterdam: Elsevier 1997) pp. 1162–1165.
14. Maj.Gen. Oleg Kalugin, former deputy KGB chief at the Russian Embassy in Washington D.C., as quoted by Bamford (note 8) p. 277.
15. *United States of America vs. Jerry Alfred Whitworth*, judgment of the U.S. Court of Appeals for the 9th Circuit, 856 *Federal Reporter* 2nd Series 1268 (November 13, 1987). Whitworth (who had previously served as a radioman at Diego Garcia in 1973–1974) was part of an espionage ring set up by U.S. Navy communications specialist John Walker (*aka* James Harper); see judgment at pp. 1271–1272, and Bamford (note 8) *ibid*.
16. See (Lt. Col.) W.C. Jeas and R. Anctil, "The Ground-Based Electro-Optical Deep Space Surveillance (GEODSS) System," *Military Electronics/Countermeasures* vol. 7 No. 11 (1981) pp. 47–51. GEODSS combines the use of telescopes, low-light TV cameras, and computers; see Edis (Chapter 1, note 3) p. 93; Bandjunis (Chapter 1, note 6) p. 246; and Harkavy (Chapter 3, note 2) p. 274. On early rivalries between the U.S. Air Force and the Central Intelligence Agency over the use of electro-optical imagery and satellite technology ("space reconnaissance wars"), see J.T. Richelson, *The Wizards of Langley: Inside the CIA's Directorate of Science and Technology* (Boulder: Westview 2002) at pp. 103–130.
17. Comprehensive Nuclear Test Ban Treaty (CTBT), opened for signature at New York on September 24, 1996, *International Legal Materials* vol. 35 (1996) p. 1439, ratified b 148 countries to date (including the UK on 6 April 1998, but not including the USA), not yet in force; see D.S. Jonas, "The Comprehensive Nuclear Test Ban Treaty: Current Legal Status in the United States and the Implications of a Nuclear Test Explosion," *New York University Journal of International Law and Politics* vol. 39 (2007) pp. 1007–1046. The United Kingdom has registered Diego Garcia as BIOT/Chagos Archipelago hydroacoustic station HA-08, and as infrasound station IS-52, under Section A(1) of the Protocol to the CTBT; see also the written answer by the UK foreign secretary to parliamentary questions on July 21, 2004, *House of Commons Debates* vol. 424 col. 340W, *British Yearbook of International Law* vol. 75 (2004) p. 672. Data collected are transmitted to the CTBT International Data Centre in Vienna/Austria; see J.K. Hettling, *Satellite Imagery for Verification and Enforcement of Public International Law* (Cologne: Heymanns 2008) at p. 121. The 1999 UK-U.S. exchange of notes also provided for the future addition of radionuclide monitoring facilities at Diego Garcia (appendix VIII, p. **115**, para. 12), which has since been done (IMS radionuclide

NOTES—CHAPTER 4

station RN-66). On the participation of the HA-08 hydrophone station in the 2001–2003 underwater sound charge experiments, see note 50.
18. See J. Berger and P. Davis [Scripps Institution of Oceanography and University of California at San Diego, part of the Incorporated Research Institutions for Seismology (IRIS) Consortium], *The IDA Network: An Element of the IRIS Global Seismographic Network*, Final Report 2001–2006 (La Jolla/CA, 22 February 2007); and note 21.
19. See C.R.C. Sheppard, "Effects of the Tsunami in the Chagos Archipelago," in D.R. Stoddart (ed.), *Tsunamis and Coral Reefs*, Atoll Research Bulletin No. 544 (Washington D.C.: Smithsonian Institution 2007) pp. 135–148.
20. See comments by U.S. Senator Olympia J. Snowe, *Boston Globe* (Boston/MA, December 29, 2004); R. Norton-Taylor, "Bulletins Sent to Diego Garcia 'Could Have Saved Lives'," *Guardian* (London, January 7, 2005); Pilger (Chapter 1, note 7) p. 60; and M. Chossudovsky, "Indian Ocean Tsunami: Why Did the Information Not Get Out?," *Global Research* (Montreal: Centre for Research on Globalization <http://www.globalresearch.ca> February 7, 2005). The U.S. State Department press release of January 14, 2005 (*Media Release* No. 660943) points out that the alert bulletins sent out by the NOAA *Pacific Tsunami Warning Center* (PTWC) in Hawaii—also to the Navy Meteorology and Oceanography Command Detachment on Diego Garcia (via the Navy's Pacific Command)—had initially misclassified the earthquake as being of an order of 8.0 on the Richter scale only.
21. IRIS Global Seismographic Network, International Deployment of Accelerometers (IDA) Network, Station DGAR (Diego Garcia, British Indian Ocean Territory) website <http://ida.ucsd.edu/Stations/dgar/index.html> maintained by the University of California at San Diego (note 18); table "special events" (December 26, 2004, updated January 2005). See also J.A. Hanson and J.R. Bowman, "Dispersive and Reflected Tsunami Signals from the 2004 Indian Ocean Tsunami Observed on Hydrophones and Seismic Stations," *Geophysical Research Letters* vol. 32 (2005) L 17606; J.A. Hanson et al., "High-Frequency Tsunami Signals of the Great Indonesian Earthquakes of 26 December 2004 and 28 March 2005," *Bulletin of the Seismographic Society of America* vol. 97 No. 1A (2007) pp. S232–S248; and Emile A. Okal et al., "Quantification of Hydrophone Records of the 2004 Sumatra Tsunami," in K. Satake et al. (eds.), *Tsunami and Its Hazards in the Indian and Pacific Oceans* (Basel: Birkhäuser 2007) pp. 309–323.
22. Chapter 1 (note 63); see M. Nowak and E. McArthur, *The United Nations Convention Against Torture: A Commentary* (Oxford: Oxford University Press 2008) p. 196; D. Weissbrodt and A. Bergquist, "Extraordinary Rendition and the Torture Convention," *Virginia Journal of International Law* vol. 46 (2006) pp. 585–650; and J. Santos Vara, "Extraordinary Renditions: The Interstate Transfer of Terrorist Suspects without Human Rights Limits," in M.J. Glennon and S. Sur (eds.), *Terrorisme et Droit International / Terrorism and International Law* (Leiden: Nijhoff/Hague Academy of International Law 2008) pp. 551–583, at 564–572. See also the Venice Commission's "Opinion on the International Legal Obligations of Council of Europe Member States in Respect of Secret Detention Facilities and Inter-State Transport of Prisoners," adopted on May 18, 2006, Opinion No. 363/2005, CDL/077 (Strasbourg: Council of Europe 2006); and note 28.
23. GAO Report (Chapter 3, note 11) p. 53 (appendix IV, also including an "internal security/dog kennel" completed by the Naval Construction Force on June 30, 1983; the report does not list professional equipment, such as waterboards).
24. Cost reimbursement, indefinite-delivery/indefinite-quantity construction contract No. N62470-04-D-4017, *Defense Industry Daily* (online June 17, 2005); cf. Chapter 3 (note 8).
25. *House of Commons Debates* vol. 422 (June 21, 2004) col. 1222W; see the submission dated October 18, 2007 from [Human Rights NGO] *Reprieve* to the House of Commons Foreign Affairs Select Committee, "Enforced Disappearance, Illegal Interstate Transfer, and Other Human Rights Abuses involving the UK Overseas Territories," in *Overseas Territories: Seventh Report 2007–08* (Chapter 1, note 64) pp. Ev203–219, at p. 207. There is a fictional

but accurate description of the British jail at Diego Garcia in the thriller by M. Dobbs, *The Lords' Day* (London: Headline Review 2007), at pp. 324–326.
26. *House of Commons Debates* vol. 424 (July 20, 2004) col. 172W; *British Yearbook of International Law* vol. 75 (2004) p. 671. According to UK foreign secretary David Miliband, the U.S. detention facility on Diego Garcia (for U.S. servicemen within the base) was decommissioned in August 2007, leaving only the jail operated by the UK authorities on the island; *House of Commons Debates* vol. 483 (November 20, 2008) col. 748W.
27. On *MSNBC Tonight* (May 6, 2004) and on *National Public Radio* (December 5, 2006); as quoted in the Reprieve Submission (note 25), footnotes 12 and 13. See C. Stafford Smith, *Bad Men: Guantánamo Bay and the Secret Prisons* (London: Weidenfeld & Nicolson 2007) p. 248.
28. See J.R. Crook, "Contemporary Practice of the United States: Continuing Controversy Regarding Secret U.S. Rendition and Detention Practices," *American Journal of International Law* vol. 100 (2006) pp. 214–247; D. Marty, *Alleged Secret Detentions and Unlawful Interstate Transfers of Detainees involving Council of Europe Member States*, 1st Report (Strasbourg: Council of Europe, June 12, 2006), 2nd Report and Explanatory Memorandum (June 7, 2007); and M. Hakimi, "The Council of Europe Addresses CIA Rendition and Detention Program," *American Journal of International Law* vol. 101 (2007) pp. 442–452.
29. Written answer by the Parliamentary Undersecretary, FCO, *House of Lords Debates* vol. 642, col. 1019–1020; *British Yearbook of International Law* vol. 74 (2003) pp. 686, 857.
30. Written answer by the Secretary of State, FCO (see note 25).
31. Written answer by the Parliamentary Undersecretary, FCO, *House of Commons Debates* vol. 426, col. 225W; *British Yearbook of International Law* vol. 75 (2004) p. 716.
32. Written answer by the Secretary of State, FCO, *House of Commons Debates* vol. 450, col. 2076W; *British Yearbook of International Law* vol. 77 (2006) p. 639; and *House of Commons Debates* vol. 464, col. 703W, *British Yearbook of International Law* vol. 78 (2007) p. 696 (now referring to the U.S./UK Political-Military Talks held in Washington D.C. on September 11 and 13, 2007).
33. Written answer by the Minister of State, FCO, *House of Lords Debates* vol. 694, col. WA25. See also the report of [non-governmental organization] *Amnesty International* for the year 2007, according to which the UK authorities told Amnesty that the UK "does not routinely keep records of flights in and out of Diego Garcia," but that they were "satisfied with [the] assurance" given by the United States that they "do not use Diego Garcia for any rendition operations"; *Amnesty International Report 2008: The State of the World's Human Rights* (London: Amnesty International 2008) p. 315.
34. "Terrorist Suspects (Renditions)," Statement by the Secretary of State, FCO, *House of Commons Debates* vol. 472, col. 547; and CIA Press Release, "Director's Statement on the Past Use of Diego Garcia" (Langley/VA, online February 21, 2008). On July 3, 2008, the Secretary of State assured Parliament that "our US allies are agreed on the need to seek our permission for any *future* renditions through UK territory" [emphasis supplied]; *House of Commons Debates* vol. 478, col. 58WS. At its Geneva session on July 7–25, 2008, the UN Human Rights Committee called on the UK government to undertake an investigation of the matter; report in Chapter 1 (note 64) p. 3, para. 13.
35. The flightlog of Gulfstream-V turbojet N379P—arriving in Diego Garcia from Athens/Greece on September 13, 2002, and continuing to Rabat/Morocco via Cairo on September 15, 2002—has been traced back to the CIA's Aero Contractors Ltd., known to have carried out hundreds of "rendition flights" worldwide since 2001; see S. Grey, *Ghost Plane: The True Story of the CIA Torture Program* (New York: St. Martin's Press 2006) p. 290, and the submission by Reprieve (note 25) p. 216. On the international legal status of the CIA flights, see M. Milde, "'Rendition Flights' and International Air Law," *Zeitschrift für Luft- und Weltraumrecht / German Journal of Air and Space Law* vol. 57 (2008) pp. 477–486.

Notes—Chapter 4

36. Submission by Reprieve (note 25) p. 208, naming *inter alia* the amphibious assault ship USS *Bataan*, known to have been stationed off Diego Garcia at the relevant times, and the naval supply ship USNS *Stockham* (T-AK 3017, formerly USNS *Soderman* T-AKR 299, originally built in Denmark as commercial container ship *Lica-Maersk*) anchored in the Diego Garcia lagoon since 2001; see also D. Campbell and R. Norton-Taylor, "US Accused of Holding Terror Suspects on Prison Ships," *Guardian* (London, online June 2, 2008). Several persons alleged to have been detained on or off Diego Garcia are identified in the submission by Reprieve *ibid*. at p. 206, and in the additional submission dated February 26, 2008, by [Reprieve Director] C.A. Stafford Smith, "Renditions and Secret Imprisonment in Diego Garcia," *ibid*. at pp. 305–307. See also the submissions dated October 15, 2007, and February 27, 2008, by Andrew Tyrie MP, Chairman of the All Party Parliamentary Group on Extraordinary Rendition, *ibid*. pp. 182–184 and 308–310; and [NGO] Redress, *The United Kingdom, Torture and Anti-Terrorism: Where the Problems Lie* (London: Redress Trust, December 2008) pp. 26–28.
37. Minutes of oral evidence taken on March 26, 2008 before the House of Commons Foreign Affairs Committee, HC 147-v (London: HMSO 2008); Report on Overseas Territories (Chapter 2, note 44) pp. 29–30. See also the oral evidence by Lord Mark Malloch-Brown KCMG PC (FCO Minister of State for Africa, Asia and the United Nations) before the Committee on May 7, 2008 (HC 2007–08) p. 533-ii (Q59); and Chapter 1 (notes 75–77). In June 2008, the issue was raised in bilateral discussions between UK Prime Minister Gordon Brown and Mauritian Prime Minister Dr. Navinchandra Ramgoolam; see Chapter 1 (note 41).
38. A. Zagorin, "Source: US Used UK Isle to Interrogate," *Time Magazine / CNN* (online July 31, 2008); A. Sparrow, "David Miliband 'Duped' Over US Rendition," *Guardian* (London, online August 1, 2008); B. Roberts, "UK Duped by US on Torture," *Daily Mirror* (London, online August 2, 2008); J. Doward, "US 'Held Suspects on British Territory in 2006'," *Observer* (International Edition, London, August 3, 2008) p. 27.
39. See the written answer by Secretary of State David W. Miliband, FCO, *House of Commons Debates* vol. 473 (March 20, 2008) col. 1288W (referring to a Washington D.C. meeting in September 2007).
40. See R. Verkaik, "Freedom of Information: Government Blocks Access to Secret Military Papers on Diego Garcia," *Independent* (London, online February 1, 2008); and generally T. Christakis, "L'État avant le droit? L'exception de 'sécurité nationale' en droit international," *Revue Générale de Droit International Public* vol. 112 (2008) pp. 5–47, at p. 38. Since 2001, UK-U.S. supplementary agreements for "infrastructure upgrades" on Diego Garcia have been concluded by simple (unpublished) "exchange of letters"; see Chapter 1 (note 28).
41. The US Freedom of Information Act of July 4, 1966 (Public Law 89-554, as amended by Public Laws 90-23, 107-306 and 110-175, 5 U.S. Code 552) has become a model for national legislation world-wide, hailed by foreign politicians as "the new Magna Carta of ecological democracy"; see J. Fischer, *Der Umbau der Industriegesellschaft: Plädoyer wider die herrschende Umweltlüge* (Frankfurt: Eichborn 1989) p. 152, English transl. in U. Beck (ed.), *Ecological Enlightenment: Essays on the Politics of Risk Society* (Atlantic Highlands/NJ: Humanity Press 1995); see also A. Roberts, "Structural Pluralism and the Right to Information," *University of Toronto Law Journal* vol. 51 (2001) pp. 243–271.
42. E.g., see T. Blanton, "The World's Right to Know," *Foreign Policy* No. 131 (2002) pp. 50–58; P. McDermott, "Withhold and Control: Information in the Bush Administration," *Kansas Journal of Law and Public Policy* vol. 12 (2002) pp. 671–692; B. Pack, "FOIA Frustration: Access to Government Documents Under the Bush Administration," *Arizona Law Review* vol. 46 (2004) pp. 815–842; S. Dykus, "Osama's Submarine: National Security and Environmental Protection After 9/11," *William and Mary Environmental Law and Policy Review* vol. 30 (2005) pp. 1–54; and N.A. Robinson, "Terrorism's Unintended Casualties:

Implications for Environmental Law in the USA and Abroad," *Environmental Policy and Law* vol. 37 (2007) pp. 125–139.

43. E.g., at the 2003 Governing Council of the United Nations Environment Programme (UNEP) in Nairobi, the U.S. delegation—together with China and the Group of 77—opposed proposals for global guidelines on access to environmental information; and at the 2004 Congress of the World Conservation Union (IUCN) in Bangkok, the U.S. delegation (upon instructions from the State Department) was alone in abstaining from a unanimous recommendation along the same lines. See also P.H. Sand, "Information Disclosure as an Instrument of Environmental Governance," *Zeitschrift für ausländisches öffentliches Recht und Völkerrecht / Heidelberg Journal of International Law* vol. 63 (2003) pp. 487–502, at p. 501.

44. 2161 United Nations Treaty Series 447, ratified by the UK on February 23, 2005. The United States did not ratify the treaty, and in November 2002 withdrew from negotiations for a follow-up UN Protocol on Pollutant Release and Transfer Registers (which was actually modeled after the 1986 US Emergency Planning and Community Right-to-Know Act, 42 U.S. Code 11001), declaring that it would no longer participate in a negotiating capacity but would "continue to follow this and other international processes dealing with the issue of PRTRs" (UN Doc. ECE/MP.PP/AC.1/2002/2, paragraph 19).

45. E-mail message to the author dated November 26, 2008 from BIOT administrator Joanne Yeadon, adding that "... as this position is unlikely to change in the foreseeable future, there are no plans to enact legislation or ratify in respect of BIOT"; *cf.* Chapter 1 (notes 60 and 64), and Chapter 2 (notes 18 and 69–75). According to an explanatory memorandum on the Aarhus Convention presented to Parliament by the Foreign and Commonwealth Office in January 2005 (Command Paper 4736), "the Overseas Territories and Crown Dependencies were consulted about the extension of the Convention to them. Whilst none are currently in a position to ratify alongside the UK, some will return to their consideration of this issue at a future opportunity" (para. 20). The UK Department for Environment, Food and Rural Affairs (DEFRA) states that the public authorities of overseas territories "are not covered by the UK Environmental Information Regulations (EIRs)" adopted under the Aarhus Convention; <http://www.defra.gov.uk/corporate/opengov/eir/faq.htm> (last modified February 15, 2007). The ten-year implementation report submitted by DEFRA to the 3rd meeting of the Parties to the Aarhus Convention (Riga, June 11–13, 2008) makes no reference to overseas territories other than Gibraltar; see UN Doc. ECE/MP.PP/IR/2008/GBR (June 6, 2008).

46. See notes 7, 11, and 17.

47. See D.K. Blackman *et al.*, "Indian Ocean Calibration Tests: Cape Town—Cocos Keeling 2003," *Proceedings of the 25th Seismic Research Review* (2003) pp. 517–523, at p. 518.

48. E.g., see C.J. Stone, *The Effects of Seismic Activity on Marine Mammals in UK Waters 1998–2000* (Aberdeen: Joint Nature Conservation Committee 2003); J.M. Van Dyke *et al.*, "Whales, Submarines and Active Sonar," *Ocean Yearbook* vol. 18 (2004) pp. 330–363; E. McCarthy, *International Regulation of Underwater Sound: Establishing Rules and Standards to Address Ocean Noise Pollution* (Boston: Kluwer Academic Publishers 2004); K.N. Scott, "International Regulation of Undersea Noise," *International and Comparative Law Quarterly* vol. 53 (2004) pp. 287–324; J. Firestone and C. Jarvis, "Response and Responsibility: Regulating Noise Pollution in the Marine Environment," *Journal of International Wildlife Law and Policy* vol. 10 (2007) pp. 109–152; J.R. Reynolds, "Submarines, Sonar, and the Death of Whales: Enforcing the Delicate Balance of Environmental Compliance and National Security in Military Training," *William and Mary Environmental Law and Policy Review* vol. 32 (2008) pp. 759–802; and I. Boyd (ed.), *The Effects of Anthropogenic Sound on Marine Mammals: A Draft Research Strategy* (Strasbourg: European Science Foundation 2008).

49. E.g., see U.S. National Research Council, *Marine Mammal Populations and Ocean Noise: Determining When Noise Causes Biologically Significant Effects* (Washington D.C.: National Academy Press 2005); M. Jasny *et al.* (eds.), *Sounding the Depths II: The Rising Toll of Sonar,*

Shipping, and Industrial Ocean Noise on Marine Life (New York: Natural Resources Defense Council 2005); T.M. Cox et al., "Understanding the Impacts of Anthropogenic Sound on Beaked Whales," *Journal of Cetacean Research and Management* vol. 7 (2006) pp. 177–187; E.C.M. Parsons et al., 'Navy Sonar and Cetaceans: Just How Much Does the Gun Need to Smoke Before We Act?', *Marine Pollution Bulletin* vol. 56 (2008) pp. 1248–1257; and Di Mento (note 11) pp. 294–373 (describing the role of the U.S. Navy as that of "a powerful bull in a china shop").

50. See S. Gray, "Giant Marine Park Plan for Chagos," *Independent* (London, online February 9, 2009); *cf.* J.E. Heyning, "Cuvier's Beaked Whale *Ziphius cavirostris*," in W.F. Perrin et al. (eds.), *Encyclopedia of Marine Mammals* (San Diego/CA: Academic Press 2002) pp. 305–307; and B.L. Taylor et al., "*Ziphius cavirostris*," in *IUCN Red List of Threatened Species* (Cambridge: IUCN/UK 2008) <http://www.iucnredlist.org>. E.g., on July 9, 2002, the research vessel *Odyssey* (Ocean Alliance, Lincoln/MA) identified a skull of Cuvier's beaked whale on Takamaka Island (in the Salomon Islands, Chagos Archipelago) and delivered it to the BIOT authorities in Diego Garcia. There have also been reports of strandings of Cuvier's beaked whales off the Maldives; see L.T. Ballance et al., "Cetacean Sightings Around the Republic of the Maldives," *Journal of Cetacean Research and Management* vol. 3 (2001) pp. 213–218.

51. Convention on the Conservation of Migratory Species of Wild Animals (CMS, 1459 *United Nations Treaty Series* 36), adopted at Bonn on June 23, 1979 (in force November 1, 1983), currently 109 member countries including the UK and Mauritius, though not the USA; the UK ratification was expressly extended to BIOT by *note verbale* dated July 17, 1985. Resolution 9.19 adopted by the 9th Conference of the Parties to CMS (held at Rome, December 1–5, 2008)

> calls on Parties and invites non-Parties whenever possible to adopt mitigation measures on the use of high intensity active naval sonars until a transparent assessment of their environmental impact on marine mammals, fish and other marine life has been completed, and as far as possible aim to prevent impacts from the use of such sonars, especially in areas known or suspected to be important habitat to species particularly sensitive to active sonars (e.g. beaked whales) and in particular where risks to marine mammals cannot be excluded, taking account of existing national measures and related research in this field.

See also Resolution 3.068 on Undersea Noise Pollution, adopted by the 3rd World Conservation Congress of the International Union for Conservation of Nature and Natural Resources (IUCN), held at Bangkok, November 17–25, 2004.

52. See Edis (Chapter 1, note 3) p. 13. A survey commissioned by the U.S. Navy in 2004 reported sightings of a humpback whale and a pilot whale; see K.J.P. Deslarzes et al., *Marine Biological Survey at United States Navy Support Facility, Diego Garcia, British Indian Ocean Territory: Final Report*, Annex O of INRMP 2005 (Chapter 5, note 41) at p. 4–10.

53. International Convention for the Regulation of Whaling (161 United Nations Treaty Series 72), adopted at Washington D.C. on December 2, 1946 (in force November 10, 1948), currently 84 members including the UK and the USA; according to the Chagos Conservation Management Plan (Chapter 5, note 59, p. 14), the convention applies to BIOT. See S. Leatherwood and G.P. Donovan (eds.), *Cetaceans and Cetacean Research in the Indian Ocean Sanctuary*, Marine Mammals Technical Report No. 3 (Nairobi: United Nations Environment Programme 1991); and M.N. de Boer et al. (eds.), *Cetaceans in the Indian Ocean Sanctuary: A Review* (Wiltshire: Whale and Dolphin Conservation Society 2003). The scientific committee of the International Whaling Commission (IWC) has concluded that "there is now compelling evidence implicating military sonar as a direct impact on beaked whales in particular"; see IWC Report 56 (2004) Annex K, para. 6.4.

54. In contrast, compare the detailed assessment under the U.S. Marine Mammal Protection Act (16 U.S. Code 1361), the Endangered Species Act (7 U.S. Code 136, 16 U.S. Code 1531), and Executive Order No. 12114 (see Chapter 5, note 38), by W. Cross and A. Hunter

(eds.), *Environmental Assessment of a Planned Low-Energy Marine Seismic Survey by the Scripps Institution of Oceanography in the Northeast Indian Ocean, May-August 2007* (King City, Ontario/Canada: LGL, January 4, 2007) pp. 13–38 (section on marine mammals); authorization for the survey (part of the "Ninety East Ridge Expedition" of R/V *Roger Revelle*, about 1,200 miles east of Diego Garcia) issued on June 20, 2007, U.S. Federal Register vol. 72 No. 127 (July 3, 2007) p. 36435.

5 Nemesis: Natural Heritage Dredged—and Drowned

1. Edis (Chapter 1, note 3) p. 13; J.M.W. Topp, "British Indian Ocean Territory: An Almost Pristine Coral Ecosystem," *Ecos* vol. 1 No. 1 (1998) pp. 27–30, and UK Foreign and Commonwealth Office, *Partnership for Progress and Prosperity: Britain and the Overseas Territories*, Cm. 424 (London: FCO 1999) pp. 50–51 ("near-pristine coral reefs").
2. Edis *ibid.* at pp. 98 and 100. See G.C. Bourne, "The Atoll of Diego Garcia and the Coral Formations of the Indian Ocean," *Proceedings of the Royal Society of London* vol. 43 (1888) pp. 440–461; Stoddart and Taylor (Chapter 1, note 12); A.R. Emery and R. Winterbottom, "A Dying Paradise," *Rotunda* vol. 12 (Toronto 1971) pp. 2–10; J.M.W. Topp, "The British Indian Ocean Territory: A Natural Paradise," *Sanctuary* vol. 26 (1997) pp. 36–39; C.R.C. Sheppard, "Corals of Chagos, and the Biogeographical Role of Chagos in the Indian Ocean," in C.R.C. Sheppard and M.R.D. Seaward (eds.), *Ecology of the Chagos Archipelago* (London: Westbury 1999) pp. 53–66; and F. Pope, "Protect This Jewel in the Indian Ocean's Crown," *Times* No. 69540 (London, 23 January 2009) p. 32. See also Sheppard, note 29; Sheppard and Spalding, note 59; and the submissions dated October 10, 2007, and April 13, 2008, by the Chagos Conservation Trust [environmental NGO] to the House of Commons Select Committee on Foreign Affairs, *Overseas Territories: Seventh Report of Session 2007–08* (Chapter 1, note 64) pp. Ev95–97 and 354–355, at p. 354: "BIOT is considered to have the most pristine tropical marine environment surviving on the planet and to be by far the richest area of marine biodiversity of the United Kingdom and its Overseas Territories." See, however, notes 8 and 68 on the FCO veto against application of the Convention on Biological Diversity to BIOT.
3. BIOT Administration, Diego Garcia Conservation (Restricted Area) Ordinance No. 6 of 1994 (in force November 24, 1997).
4. Convention for the Protection of the World Cultural and Natural Heritage, adopted under UNESCO auspices at Paris on November 16, 1972 (in force December 17, 1975, 1037 United Nations Treaty Series 151), currently 185 member states; ratified by the United Kingdom on May 25, 1984 (with a declaration extending it to all British overseas territories *except* BIOT), by the United States on December 7, 1973, and by Mauritius on September 19, 1995.
5. BIOT administration, *The British Indian Ocean Territory Conservation Policy* (London: FCO 1997) p. 1; see P. Bridgewater and I. Orr, in Proceedings of the 2007 London Conference of the Chagos Conservation Trust, "The Future Conservation of the Chagos," *Chagos News* [2008] No. 31, at pp. 6 and 13. So far, however, neither the United States nor Mauritius appear to have been consulted with regard to formal UNESCO nomination of the Chagos as a World Heritage site. On the other hand, the non-governmental Mauritian Marine Conservation Society (MMCS) has since 1996 called for designation of the Chagos Archipelago as a (Mauritian) world heritage site; see Ollivry (Chapter 1, note 39) at p. 30.
6. D. Proctor and L.V. Fleming (eds.), *Biodiversity: The UK Overseas Territories* (Peterborough: Joint Nature Conservation Committee 1999) at p. 38; reiterated in H. Berends and J. Pears, *Overseas Countries and Territories: Final Report to the European Commission/Europe Aid Cooperation Office*, Part 2 Section B: Indian Ocean Region (Brussels: NIRAS/PINSISI Consultants, January 2007) p. 36. According to a census of seabirds conducted by the Royal

Notes—Chapter 5

Naval Birdwatching Society (RNBWS) in the southern part of Diego Garcia in November 2007 (with funding from the UK Overseas Territories Environment Programme, jointly sponsored by the FCO and the Department for International Development), "this previously inhabited part of the atoll was left to nature in the early 1960s and made out of bounds to personnel serving on Diego Garcia in the early 1970s. Possibly as a result of this lack of disturbance, seabirds have re-colonised this part of Diego Garcia, after being absent for the better part of a century"; [Maj.] P. Carr, *Forum News* (UK Overseas Territories Conservation Forum, July 2008) p. 15; *Chagos News* No. 32 (2008) p. 12.

7. "They really wanted to pretend it did not exist, in case questions were asked": Prof. Charles R.C. Sheppard FLS [Conservation Consultant to the BIOT Administration], as quoted by F. Pearce, "Paradise Lost to Pirates," *New Scientist* vol. 156 No. 2104 (October 18, 1997) p. 15.

8. Convention on Biological Diversity, adopted at Rio de Janeiro on June 5, 1992 (in force December 29, 1993), 1760 United Nations Treaty Series 79; ratified by Mauritius on September 4, 1992 (Chapter 1, note 43), and by the United Kingdom on June 3, 1994, also on behalf of selected overseas territories (*not* including the BIOT; see note 68 on the FCO veto). The United States signed the convention on June 4, 1993, but never ratified it. Mauritius has made it clear that its ratification applies to the Chagos Archipelago and Diego Garcia; see National Parks and Conservation Service, *First National Report to the Convention on Biological Diversity* (Port Louis: Ministry of Agriculture, Food Technology and Natural Resources, November 2000) pp. 8 and 27.

9. Convention on Wetlands of International Importance Especially as Waterfowl Habitat, adopted at Ramsar/Iran on February 2, 1971 (in force December 21, 1975, amended in 1982 and 1987), 996 United Nations Treaty Series 245, currently 158 member states; ratified by the United Kingdom with effect from May 5, 1976 (with a declaration extending the convention to British overseas territories, though *not including BIOT*), by the United States with effect from April 18, 1987, and by Mauritius with effect from September 30, 2001.

10. Diego Garcia is listed as site No. 1077 (*2UK001*, with a total area of 135.4 square miles = 354.2 square kilometers); see *Ramsar Bulletin Board* (September 4, 2001); D. Taylor *et al.* (eds.), *Ramsar Sites Directory and Overview*, 8th edn. (Wageningen: Wetlands International 2005); and M.W. Pienkowski, *Review of Existing and Proposed Ramsar Sites in UK Overseas Territories and Crown Dependencies* (London: Department of Environment, Food and Rural Affairs 2005) pp. 98–101, with a map at p. 865 (see Map 3, p. **60**).

11. See note 26, p. 12.

12. See "NAVFAC Far East Provides Integral Environmental Support," U.S. secretary of defense Press Release (Yokosuka/Japan, March 3, 2008). The coconut crab (*Birgus latro*) is a protected wildlife species on Diego Garcia.

13. Bandjunis (Chapter 1, note 6) p. 49; see also Menzie *et al.* (note 28) p. 1.

14. Richard Edis CMG [later British Ambassador to Mozambique, Tunisia and Algeria] (Chapter 1, note 3) at p. 86. The base was nick-named "Dodge City," after the American TV western series, *Gunsmoke*.

15. Edis *ibid.* p. 87.

16. As told in the "war stories" by Diego Garcia veterans on the website of T. Morris (Chapter 3, note 18): R. Lehner (NMCB-1/Explosive Ordnance Detachment, 1977–1978); see also the photographs in T. Grenier, "Blowing Holes in the Reef for Fun and Profit" (*ibid.*, 1974, posted April 17, 2004), and Y. Levi (an Israeli frogman from *Oceaneering International*), "Blowing Open the Ship Channel" (*ibid.*, 1974, updated April 30, 2007).

17. See Chapter 3, notes 6–7.

18. See Chapter 3, notes 8–11.

19. GAO report (Chapter 3, note 11) at pp. 46 (App. I) and 50 (App. II).

20. 173,000 U.S. tons; see Tucker and Doughty (Chapter 3, note 10) at pp. 202–207, and note 46.

21. Integrated Natural Resources Management Plan Diego Garcia 2005 (note 41) p. 3–4. Part of the dredging took place inside the area designated since 2001 as an internationally protected wetland under the Ramsar Convention (note 10).
22. Tucker and Doughty (Chapter 3, note 10) at p. 213.
23. See Chapter 3, note 19; Rais (Chapter 1, note 35) at pp. 84–85; J. Fuller, "Dateline Diego Garcia: Paved-Over Paradise," *Foreign Policy* No. 28 (Autumn 1977) pp. 175–186; and M. Calabresi, "Paradise in Concrete: Postcard from Diego Garcia," *Time Magazine* (online September 6, 2007).
24. Edis (Chapter 1, note 3) at pp. 3–4.
25. Naval Communications Facility on Diego Garcia: Agreement Supplementing the Agreement of December 30, 1966, October 24, 1972 (appendix III, p. **86**), Article 7, which also provides for the future restoration of the three islets at the mouth of the lagoon. An identical clause—"feeble as it was," in the words of Edis (Chapter 1, note 3) p. 99—was inserted in the 1976 Supplement (appendix IV, p. **94**) as Art. 8.
26. U.S. Naval Facilities Engineering Command/Pacific Division, *Natural Resources Conservation Land Management Plan for Diego Garcia, British Indian Ocean Territories* (Pearl Harbor/HI, December 1973), prepared by a Navy Natural Resources Specialist, on the basis of a field mission from August 6 to 18, 1973. Document released by the Department of the Navy on May 21, 2008 (upon appeal to the Office of the General Counsel), under Freedom of Information Act (FOIA) File 5720#08–29.
27. U.S. Naval Facilities Engineering Command/Pacific Division, *Natural Resources Management Plan for Diego Garcia, Indian Ocean (British Indian Ocean Territory)*, (Pearl Harbor/HI, October 25, 1981), prepared by a Navy Natural Resources Team, on the basis of a field mission in February 1981. Document released under FOIA File 7720#08–29 (note 26).
28. C. Menzie et al., *Environmental Survey of Construction and Dredging Related Activities on Diego Garcia, Indian Ocean* (Waltham/MA: EG&G Environmental Consultants, July 1980), prepared as part of a topographic and hydrographic survey of Diego Garcia conducted by *EG&G Environmental Consultants*/Singapore, on the basis of a field mission in May 1980. Document released under FOIA File 7720#08-29 (note 26).
29. *Ibid.* pp. 90–92. The report also cautioned against beach erosion risks caused by the massive destruction of protective *Scaevola* ("scavvy") vegetation, and recommended replanting (pp. 37 and 91); along the same lines, see C.R.C. Sheppard, "The Chagos Archipelago," in C.R.C. Sheppard (ed.), *Seas at the Millennium: An Environmental Evaluation*, vol. 2 (Amsterdam: Pergamon 2000) pp. 221–232, at p. 230.
30. Menzie et al. (note 28) p. 89.
31. Sheppard (note 29) p. 230, and Chapter 3 (note 17).
32. Sheppard *ibid.*; Sheppard and Spalding (note 59) at p. 30.
33. See appendix IV, Supplementary Arrangements 1976 for Diego Garcia Facility, Related Note No. 5, item #1 (p. **103**). The arrangements were signed on February 25, 1976 by Admiral David H. Bagley (commander-in-chief, U.S. Naval Forces in Europe), and Vice Admiral Raymond D. Lygo (later Sir Raymond Lygo KCB, chairman of Royal Ordnance PLC, now BAE Systems Land & Armaments LP, largest defense contractor in Europe).
34. See pp. **35** and **53**.
35. Naval Support Facility on Diego Garcia: Agreement Supplementing the Agreement of December 30, 1966, December 13,1982 (appendix VI, p. **108**), para. 5. See also the written answer by the FCO Parliamentary Undersecretary (Baroness Elizabeth Symons of Vernham Dean PC), in response to questions in the House of Lords, *Lords Hansard* vol. 600 (April 26, 1999) col. WA2: "The American authorities on Diego Garcia had been involved in dumping, dredging and coral reef blasting prior to the 1982 UK/US Supplemental Agreement concerning the United States Navy Support Facility on Diego Garcia. No estimate of any damage done was made at that point."
36. See Chapter 1 (note 67).

Notes—Chapter 5

37. [Cdr.] J.M.W. Topp OBE FLS, *Annual Reports of the Conservation Advisor to BIOT Administration 1993–2002*, as quoted by Sheppard and Spalding (note 59) p. 28.
38. Executive Order No. 12114 (signed by President Carter on January 4, 1979), 44 Federal Register 1957, 3 Code of Federal Regulations 356 (1979); superseding earlier Executive Orders No. 11752 (December 19, 1973) and No. 12088 (October 13, 1978). See also the Directives and Instructions of the U.S. Department of Defense listed in note 42.
39. On the "overseas reach" of U.S. law on environmental impact assessment, see L.J. Schiffer, "The National Environmental Policy Act Today, with an Emphasis on Its Application Across U.S. Borders," *Duke Environmental Law and Policy Forum* vol. 14 (2003) pp. 325–345; and S. Dycus, *National Defense and the Environment* (Hanover/NH: University Press of New England 1996) at pp. 26–30.
40. OPNAVINST.1B: *Navy Environmental and Natural Resources Program Manual*, Appendix E (CH-4, as updated on June 4, 2003); see [Maj.] J. Rawcliffe (ed.), *Operational Law Handbook* (Charlottesville/VA: U.S. Judge Advocate General Legal Center and School 2007), Chapter 18-5.6.
41. Naval Facility Engineering Command/Pacific Division, *Natural Resources Management Plan, U.S. Naval Support Facility Diego Garcia* (Pearl Harbor/HI, April 1997, cited as NRMP Diego Garcia 1997), prepared according to OPNAVINST 5090.1B (November 1, 1994) and Final Governing Standards for Diego Garcia (October 1994) on the basis of scientific surveys undertaken in October 1995 and May 1996, updated and expanded as *Integrated Natural Resources Management Plan, Diego Garcia, British Indian Ocean Territory* (Pearl Harbor/HI, September 2005, cited as INRMP Diego Garcia 2005), prepared by U.S. Navy biologist V.E. Pepi on the basis of a survey undertaken in August 2005 (update scheduled for 2010); documents released by the Department of the Navy on July 19, 2007 (upon request filed on January 25, 2007), under Freedom of Information Act (FOIA) File CNFJ 07-26.
42. *Final Governing Standards Diego Garcia*, Chapter 17 (Environmental Effects Abroad), November 2001, issued pursuant to Department of Defense Directive DoD 6050.16 (*DoD Policy for Establishing and Implementing Environmental Standards at Overseas Installations*, September 20, 1991); see also DoD Instructions 4715.5 (*Management of Environmental Compliance at Overseas Installations*, April 22, 1996) and 4715.8 (*Environmental Remediation for DoD Activities Overseas*, February 2, 1998).
43. E.g., see the decision "to forego preparation of an Environmental Review or Environmental Study" for MILCON Project P-160 (Wharf Improvements and *SSGN* Shore Support Facilities), dated May 22, 2006 (Cdr. CEC, DG), and July 28, 2006 (Regional Environmental Program Director, COMNAVFORJAPAN): "This action does not involve non-compliance with the Final Governing Standards, OPNAVINST 5090.1B CH-4 or the host nation's environmental standards." Unpublished document released under FOIA File 5720#08-29 (note 26); see also Chapter 3 (note 47) on related further construction works initiated in 2007.
44. In response to multiple requests and appeals for documentation under the U.S. *Freedom of Information Act* (FOIA), initiated since January 2007, NAVFAC Pacific was unable to produce or locate a single environmental impact statement, review or study for any of the coral-blasting and coral-mining projects undertaken by the "Seabees" in the 1970s (Chapter 3, note 5) or under the dredging contracts awarded to Taiwanese, Japanese and Anglo-American firms in the 1980s (Chapter 3, notes 7–11); see correspondence with the author under FOIA Files CNFJ 07-26, 07-23, 07–99 (2007) and 5720#08-29 (2008). See also note 59.
45. Natural Resources Management Plan Diego Garcia 1997 (note 41) p. B-15.
46. W.A. Whistler, "Botanical Survey of Diego Garcia, Chagos Archipelago, British Indian Ocean Territory," in NRMP Diego Garcia 1997 (note 41) pp. 4–5 and 4–9, noting a 31 per cent observed increase in unintentionally introduced plant species from 1988 to 1996, imported with piles of sand from Malaysia (p. **53**). Yet, eight years later, observations of

the same species were confirmed by J. Rivers, "Botanical Survey Update of Diego Garcia, Chagos Archipelago, British Indian Ocean Territory," in INRMP Diego Garcia 2005 (note 41) p. E2–5, warning with regard to *Leucaena* in particular that "if uncontrolled, this species can completely overtake all other species creating monotypic scrub," and once again "strongly recommending" its control and eradication (p. 6–4 of the plan); *cf.* S. Lowe *et al.* (eds.), *100 of the World's Worst Invasive Alien Species: A Selection from the Global Invasive Species Database* (Auckland: IUCN/ISSG 2004) p. 6; S. Matthews and K. Brand, *Africa Invaded: The Growing Danger of Invasive Alien Species* (Nairobi: Global Invasive Species Programme 2004) p. 29; and N. Pallawatta *et al.* (eds.), *Invasive Alien Species in South-Southeast Asia: National Reports and Directory of Resources* (Nairobi: Global Invasive Species Programme 2003). See also fig. 23 (number of introduced plant species recorded on Diego Garcia, 1967–1997) in C.R.C. Sheppard *et al.*, *Science in Chagos: What We Know and What We Need To Know* (London: Chagos Conservation Trust 2009) p. 17. In contrast, see the astonishing assertion by Pope (note 2): "Diego Garcia harbour is one of the few places in the world where surveys find not a single non-native species—indicating an ecosystem without man-made cracks where non-native species can take hold."
47. NRMP Diego Garcia 1997 (note 41) p. A–22.
48. Response to the author dated July 19, 2007, by Lt.Cdr. I.C. Le Moyne Jr., Deputy Judge Advocate, COMNAVFORJAPAN, to freedom-of-information request CNFJ 07-26; see also Sheppard and Spalding (note 59) at p. 32.
49. Edis (Chapter 1, note 3) p. 90.
50. See generally C.D. Hunt Jr., "Hydrogeology of Diego Garcia," in H.L. Vacher and T. Quinn (eds.), *Geology of Carbonate Islands,* Developments in Sedimentology No. 54 (Amsterdam: Elsevier 1997) pp. 909–931; D.P. Leigh *et al.*, "Vertically Propagated Tidal Fluctuation at Atoll Island Diego Garcia BIOT," *Proceedings of the Pacific Environmental Restoration Conference* (Honolulu/HI, April 4–7, 2000); and T.A. Morris, *Ghyben-Herzberg (Fresh Water) Lenses in the Chagos* (London: Chagos Conservation Trust Factsheet, April 2008).
51. For a first-hand description see Morris (Chapter 3, note 18), section "General Ecology of Diego Garcia: Human Use and Industrial Waste," last updated March 31, 2008. Recovery was initiated in 1987 by the civilian 'base operation and support' contractor at the time, FEBROE (a joint venture of Frank E. Basil Inc., a multinational engineering consultant company with headquarters in Washington,D.C. and Athens/Greece, and Burns & Roe Services Corp. of Bedrock/CO).
52. See D.H. Kampbell *et al.*, "Natural Bioventing Remediation from Tidal Wave Action at a Field Site," in L.E. Erickson *et al.* (eds.), *Proceedings of the HSRC/WERC Joint Conference on the Environment* (Manhattan/KA: Kansas State University, May 1996); J.E. Hansen, *Cleanup Plan for Fuel Spills at Air Operations Ramp, Diego Garcia, British Indian Ocean Territory*, Pentagon Report A87824 (Brooks City/TX: U.S. Air Force Center for Environmental Excellence, May 25, 1999); and G.W. Tribble, *Groundwater Geochemistry at the South Ramp Jet Fuel Leak, Diego Garcia Atoll*, Report SuDoc I 19.42/4:01-4090 (Washington,D.C.: U.S. Geological Survey 2001).
53. Hansen (note 52) p. 15; see also D.P. Leigh *et al.*, "A Comparison of Natural Attenuation Characteristics on Two Atoll Islands," *Proceedings of the 6th International In Situ and On-Site Bioremediation Symposium, San Diego/CA, June 4–7, 2001* (Columbus/OH: Battelle Press 2001); J.E. Hansen and D. Kampbell, "Large-Scale Bioslurping Operations Used for Fuel Recovery," *EPA Technology News and Trends* Issue 8 (September 2003) pp. 2–3; and J.E. Hansen, "LNAPL Characterization and Recovery at Diego Garcia," in *Summary of the Remediation Technologies Development Forum* (San Antonio/TX: Non-Aqueous Phase Liquid Cleanup Alliance Meeting, February 7–8, 2006), Attachment H. The spills were not widely publicized, and there is no reference to them in the otherwise comprehensive worldwide *IncidentNews* website maintained by the U.S. National Oceanic and Atmospheric Administration (NOAA).

54. On a 109,000 gallon (465,000 liters) *JP-5* fuel leakage at the U.S. Navy's Arraiján base in Panama in January 1995, see S. de Grimaldo (ed.), *Bases Militares: Seguridad Ambiental y Desarrollo Nacional* (Panamá: Autoridad de la Región Interoceánica 1995) p. 103; and J. Lindsay-Poland, *Emperors in the Jungle: The Hidden History of the U.S. in Panama* (Durham/NC: Duke University Press 2003) p. 147. On a 100,000 gallon (427,000 liters) *JP-5* fuel spill at Naval Air Station Roosevelt Roads, Cieba/Puerto Rico, in October 1999, see *IncidentNews* (note 53). On jet fuel spills at Clark Air Base in the Philippines (closed down in 1991 as a result of local opposition), see T.M. Proechel, "Solving International Environmental Crimes: The International Environmental Military Base Reconstruction Act, a Proposal, and a Solution," *Loyola at Los Angeles International and Comparative Law Review* vol. 29 (2007) pp. 121–151, at pp. 127–128.
55. Report of water quality measurements taken during a 2006 research expedition; see C.R.C. Sheppard, "The Evolving Science in Chagos," *Chagos News* No. 30 (2007) pp. 2–15, at p. 13. The measurements were, however, not specifically targeted on contamination by fuel residues; see C. Guitart *et al.*, "Negligible Risks to Corals from Antifouling Booster Biocides and Triazine Herbicides in Coastal Waters of the Chagos Archipelago," *Marine Pollution Bulletin* vol. 54 (2007) pp. 226–232.
56. Edis (Chapter 1, note 3) p. 103. Note that U.S. Department of Defense Instruction 4715.8 (note 42) expressly requires remediation of all "known imminent and substantial endangerments to human health and safety due to environmental contamination that was caused by DoD operations."
57. See J. Paquin, "2004 API Awards for Excellence in Fuels Management," *Navy Supply Corps Newsletter* (November–December 2004).
58. Topp (note 37).
59. C.R.C. Sheppard and M. Spalding, *Chagos Conservation Management Plan* (London: FCO, October 2003), at p. 28. In the chapter devoted to Diego Garcia (pp. 28–34), the authors raise numerous critical questions regarding gaps in the coverage and practical implementation of the 1997 NRMP, and note the absence of "environmental impact assessments and studies done over the past 25 years prior to major works" (p. 30).
60. INRMP Diego Garcia 2005 (note 41) p. 2–1; see also Chapter 1 (note 68), and Proctor and Fleming (note 6) p. 42: "Enforcement of conservation measures, such as for the existing bird sanctuaries, is the responsibility of the senior UK representative stationed on the island in his role as Magistrate."
61. Baroness Symons of Vernham Dean (note 35). As in the case of the use of BIOT for CIA "rendition flights," of course, "it is a matter of political will of States either to act or to adopt the attitude 'see no evil' "(Milde, Chapter 4, note 35, at p. 486).
62. Written answer by the Parliamentary Undersecretary of State, FCO, *House of Commons Debates* vol. 423 (July 12, 2004) col. 963W; *British Yearbook of International Law* vol. 75 (2004) p. 951.
63. Written answer by the Parliamentary Undersecretary of State, FCO, *House of Commons Debates* vol. 424 (July 19, 2004) col. 62W; *British Yearbook of International Law* vol. 75 (2004) p. 671.
64. Notes 8–9; see the reference to U.S. ratification of the Ramsar Convention in the navy's Natural Resources Management Plan 1997 (note 41) at p. ES–3/footnote 2. The designation of Diego Garcia as a Ramsar site became effective on July 4, 2001.
65. Ramsar Convention Secretariat, *Information Sheet on Ramsar Wetlands (RIS)* UK61002, by the UK Joint Nature Conservation Committee (Version 3, June 13, 2008), at section 26. Note that the 1982 UK–U.S. Supplementary Arrangements on Diego Garcia expressly provide that "the British Government Representative will be informed of any significant fall in the level of the water table or *any significant deterioration in the quality of the water, with a view to jointly agreed remedial action*"; appendix VI, para. 3 [p. **108**, emphasis supplied].

66. E.g., see the information on Diego Garcia in the UK national report to the 9th Meeting of the Conference of the Contracting Parties to the Ramsar Convention (Uganda, 2005) p. 75, which mentions neither the U.S. fuel spills nor the massive introduction of invasive alien plant species since 1988 (see note 46), let alone radiation risks (see Chapter 3, notes 50–53), undersea noise pollution (see Chapter 4, text at notes 46–54), or plans for an offshore power plant near the Ramsar site (see note 82). There is *no* reference at all to Diego Garcia in the national report submitted by the UK Department of Environment, Food and Rural Affairs (DEFRA) to the 10th Meeting of the Conference (held in Changwon, Republic of Korea, from October 28 to November 4, 2008).
67. See note 8.
68. See F. Pearce, "Britain's Abandoned Empire," *New Scientist* vol. 142 No. 1922 (April 23, 1994) p. 26. The *BIOT Environment Charter* proclaimed by the FCO on September 26, 2001 includes a formal commitment by the UK Government to "facilitate the extension of the UK's ratification of Multilateral Environmental Agreements of benefit to the BIOT and which the BIOT has the capacity to implement." However, no additional MEAs have been extended to BIOT since 2001; see M. Pienkowski (ed.), *Measures of Performance by 2007 of UKOTs and UK Government in Implementing the 2001 Environment Charters or Their Equivalents* (Peterborough: United Kingdom Overseas Territories Conservation Forum 2007) p. 14.
69. See the section on "alien species" in: J. Williams (ed.), *United Kingdom and its Overseas Territories and Crown Dependencies: Third National Report to the Convention on Biological Diversity* (London: Department for Environment, Food and Rural Affairs 2005) pp. 76–77, which reports on invasive species control and eradication in Ascension Island, the British Virgin Islands, Cayman Islands, Falkland Islands, St. Helena and Tristan da Cunha—but *not* in BIOT. Mauritius, which also ratified the convention and considers it applicable to the Chagos Archipelago and Diego Garcia (see note 8), makes no reference to the archipelago in connection with invasive alien species either; see *Convention on Biological Diversity: Third National Report for the Republic of Mauritius* (Port Louis: Ministry of Environment and National Development Unit, October 2006) pp. 6, 48, 68, 93, 101–104.
70. As quoted by Pearce (note 68).
71. See UN Office of Legal Affairs, *Multilateral Treaties Deposited with the Secretary-General: Status as at 31 December 2006*, ST/LEG/SER/E.25 (New York: United Nations 2007).
72. See Sheppard (note 55) p. 7, figure 1: *Chagos Annual Average Sea Temperature 1871–2006*, showing a dramatic increase of warm periods from 1990 on, as compared to the standard reference period 1960–1989; *cf.* C.R.C. Sheppard *et al.*, *Science in Chagos: What We Know And What We Need To Know* (London: Chagos Conservation Trust 2009) pp. 3–5; N.J. Abram *et al.*, "Seasonal Characteristics of the Indian Ocean Dipole during the Holozene Epoch," *Nature* vol. 445 No. 7125 (January 18, 2007) pp. 299–302; and the recent upward revisions of *global* ocean temperature forecasts by C.M. Domingues *et al.*, "Improved Estimates of Upper–Ocean Warming and Multi-Decadal Sea–Level Rise," *Nature* vol. 453 No. 7198 (June 19, 2008) pp. 1090–1093.
73. C.R.C. Sheppard, "A Marine Nutcracker," *Marine Pollution Bulletin* vol. 56 (2008) pp. 799–800; see also *id.*, "Coral Decline and Weather Patterns Over 20 Years in the Chagos Archipelago, Central Indian Ocean," *Ambio* vol. 28 (1999) pp. 472–478; J. Raven *et al.* (eds.) *Ocean Acidification Due to Increasing Atmospheric Carbon Dioxide*, Policy Document 12/05 (London: Royal Society 2005); O. Høegh–Guldberg *et al.*, "Coral Reefs Under Rapid Climate Change and Ocean Acidification," *Science* vol. 318 No. 5857 (December 14, 2007) pp. 1737–1742; International Society for Reef Studies, *Coral Reefs and Ocean Acidification*, Briefing Paper No. 5 (Lawrence/KS: ISRS 2008); U. Riebesell,'Climate Change: Acid Test for Marine Biodiversity', *Nature* vol. 454 No. 7200 (July 3, 2008) pp. 46-47; D. Kline *et al.*, "Impacts of Ocean Acidification and Warming on Calcifying Coral Reef Organisms," *Proceedings of 11th International Coral Reef Symposium* (Fort Lauderdale/FL, July 7–11, 2008);

Notes—Chapter 5

and N. Graham et al., "Climate Warming and the Ocean–Scale Integrity of Coral Reef Ecosystems," *ibid.* (July 9, 2008).

74. See C.R.C. Sheppard, "Island Elevations, Reef Condition and Sea Level Rise in Atolls of Chagos, British Indian Ocean Territory," in O. Lindén et al. (eds.), *Coral Reef Degradation in the Indian Ocean* (Kalmar: IUCN/Cordio 2002) pp. 26–35; J.R. Turner, Proceedings of the 2007 Chagos Conservation Trust Conference (note 5) pp. 11–12; Turner et al. (Chapter 2, note 60) pp. 25–26; and J. Petit (ed.), *Climate Change and Biodiversity in the European Union Overseas Entities*, pre–conference document (Saint–Denis/Réunion: IUCN, July 7–11, 2008) vol. 3, pp. 90–93.

75. Forecasts of global sea-level rise are currently being revised upwards from earlier reports of the *Intergovernmental Panel on Climate Change* (IPCC); see S. Rahmstorf, "A Semi-Empirical Approach to Projecting Future Sea-Level Rise," *Science* vol. 315 No. 5810 (January 19, 2007) pp. 368–370, with technical comments by S. Holgate et al., *Science* vol. 317 No. 5846 (28 September 2007) p. 1866; and W.T. Pfeffer et al., "Kinematic Constraints on Glacier Contributions to 21st-Century Sea-Level Rise," *Science* vol. 321 No. 584 (September 5, 2008) pp. 1340–1343.

76. See J.A. Church et al., "Sea-Level Rise at Tropical Pacific and Indian Ocean Islands," *Global and Planetary Change* vol. 53 (2006) pp. 155–168; and C. Bouchard, "Climate Change, Sea Level Rise, and Development in Small Island States and Territories of the Indian Ocean," in T. Doyle and M. Risely (eds.), *Crucible for Survival: Environmental Security and Justice in the Indian Ocean Region* (New Brunswick/NJ: Rutgers University Press 2008) pp. 258–272 (referring to Diego Garcia at p. 270/note 2). The newly elected president of the Maldives, Mohamed Nasheed, has announced plans for a trust fund to acquire alternative land for his country's 300,000 inhabitants elsewhere (possibly in India, Sri Lanka, or even Australia); see R. Ramesh, "Paradise Almost Lost: Maldives Seek to Buy a New Homeland," *Guardian* No. 50,442 (November 10, 2008); and J. Henley, "The Last Days of Paradise," *Guardian* (online, November 11, 2008).

77. See Chapter 2 (note 60) and Chapter 6 (note 11). Opinions among the surviving Chagossians are divided, but a core group of about 150 families is determined to return to their homeland regardless of climatic risks; see L. Jeffery, 'Victims and Patrons: Strategic Alliances and the Anti-Politics of Victimhood Among Displaced Chagossians and Their Supporters', *History and Anthropology* vol. 17 (2006) pp. 297–12; D. Vine and L. Jeffery, "Give Us Back Diego Garcia: Unity and Division Among Activists in the Indian Ocean," in Lutz (Chapter 3, note 2) pp. 181—217; and Allen (Chapter 1, note 64).

78. S. Goodman (ed.), *National Security and the Threat of Climate Change* (Alexandria/VA: Center for Naval Analyses 2007). By coincidence, among the top–level scientific advisers of the panel was Robert A. Frosch (former Assistant Executive Director of the UN Environment Programme, and now at Harvard University's Kennedy School of Government), who had first proposed—in 1970, in his capacity as Assistant Secretary of the Navy for Research and Development at the time—that the United States should establish an intelligence–collection base on Diego Garcia; see Chapter 4 (note 7).

79. Goodman (note 78) p. 48; see also C. Abbott, *An Uncertain Future: Law Enforcement, National Security and Climate Change* (London: Oxford Research Group Briefing Paper, January 2008) at p. 10.

80. Goodman *ibid.* (recommendation 5). Local shoreline erosion is currently controlled by coastal armoring and monitoring; see "Moffatt & Nichol Performs Emergency Shoreline Evaluation on Diego Garcia," Press Release by Moffatt & Nichol Engineers (Long Beach/CA, February 16, 2009).

81. See Chapter 3 (note 2).

82. Map courtesy of the Ramsar Convention Secretariat; see Pienkowski (note 10 above) p. 865. The Diego Garcia lagoon (which is part of Britain's Ramsar site No. 1077/2UK001) must be the world's only internationally registered nature reserve that also serves as habitat

to nuclear submarines, ordnance supply vessels for landmines and cluster bombs, and prison ships; see Map 2 (p. **12**) and Chapter 1 (notes 69–76).

83. See Chapter 3 (note 47). The infrastructure upgrades planned will require a further supplementary agreement with the United Kingdom, currently under negotiation between the UK Ministry of Defence and the U.S. Department of State; see Erickson *et al.* (Chapter 1, note 11) p. 18.

84. Under a joint-venture project (see pp. **36** and **109**), Ocean Engineering and Energy Systems International Inc. (OCEES) of Honolulu/HI and Ocean Thermal Energy Conversion Systems Ltd. (OTECS) of Orpington/England are designing an ocean thermal energy conversion (OTEC) facility to be built off Diego Garcia. According to the U.S. Navy's *Fiscal Year 2006 Annual Energy Management Report*, phase III of the project "is moving forward with support from the U.S. and British commands" (p. 16), and with funding in the range of $100 million to be provided by a consortium of European banks; see also the testimony of Wayne Arny, Deputy Undersecretary of Defense for Installations and Environment, before the Subcommittee on Readiness of the House of Representatives' Armed Services Committee (March 13, 2008) p. 7. The plant (which could be up and running by the end of 2011) is to supply 8 megawatts of electricity, with sufficient power to desalinate 1.25 million gallons [4.73 million liters] of seawater per day and to provide seawater air-conditioning (SWAC), so as to make the U.S. base independent of fuel supplies. See P. McKenna, "Plumbing the Oceans Could Bring Limitless Clean Energy," *New Scientist* vol. 200 No. 2683 (November 22, 2008) pp. 28–29, at 29; and T. Kauffman, "Navy Slashes Energy Use Through Variety of Projects," *Federal Times* (Springfield/VA, online December 24, 2008); *cf.* H.J. Krock, *Ocean Thermal Energy Conversion Feasibility Study for the Navy*, Report to U.S. Naval Facilities Engineering Command, Contract No. N47408–94-D-1038, D.O. 0014 (Manoa/HI: University of Hawaii 1996); and D.E. Lennard, "Ocean Thermal Energy Conversion," in J. Trinnaman and A. Clarke (eds.), *Survey of Energy Resources* (21st edn. London: World Energy Council 2007), pp. 565–581, at 572. Details of the project are confidential, according to the U.S. Navy's response to freedom-of-information request FY-09-08 (NAVFAC Far East, April 16, 2009). No comprehensive environmental impact assessments of the project appear to have been undertaken either by U.S. or British authorities so far, except for a hypothetical desk study by an engineer at the U.S. Naval Academy in Annapolis/MD; see C. Wu, "Environmental Impact and Design of an OTEC Plant," *International Journal of Global Energy Issues* vol. 15 (2001) pp. 186–194.

85. Erickson *et al.* (Chapter 1, note 11) p. 27. See also Chapter 3 (note 59); and S. Daggett *et al.*, "Fiscal Year 2008 Supplemental Appropriations for Global War on Terror Military Operations, International Affairs, and Other Purposes," *Congressional Research Service Report* RL34278 (Washington,D.C.: U.S. Government Printing Office, December 10, 2007), table 2: military construction.

86. Goodman (note 78), recommendation 2 at p. 46. *Cf.* K.M. Campbell *et al.* (eds.), *The Age of Consequences: The Foreign Policy and National Security Implications of Global Climate Change* (Washington,D.C.: Center for Strategic and International Studies 2007); J.W. Busby, *Climate Change and National Security: An Agenda for Action*, Council Special Report No. 32 (New York: Council on Foreign Relations 2007); R. Schubert *et al.* (eds.), *World in Transition: Climate Change as a Security Risk* (London: Earthscan & German Advisory Council on Climate Change 2008); and T. Fingar, *National Intelligence Assessment on the National Security Implications of Global Climate Change to 2030*, Statement to the U.S. House of Representatives Permanent Select Committee on Intelligence and Select Committee on Energy Independence and Global Warming (Washington, D.C.: Government Printing Office, June 25, 2008).

87. For a sampling of the debate on potential international legal implications, *cf.* A. Gillespie, "Small Island States in the Face of Climate Change: The End of the Line in International Environmental Responsibility," *University of California at Los Angeles Journal of Environmental*

Law and Policy vol. 22 (2004) pp. 107–129; R. Verheyen, *Climate Change Damage and International Law: Prevention Duties and State Responsibility* (Leiden: Nijhoff 2005); H.M. Osofsky, "The Geography of Climate Change Litigation: Implications for Transnational Regulatory Governance," *Washington University Law Quarterly* vol. 83 (2005) pp. 1789–1856; J. Gupta, "Legal Steps Outside the Climate Convention: Litigation as a Tool to Address Climate Change," *Review of European Community and International Environmental Law* vol. 16 (2007) pp. 76–86; T. Koivurova, "International Legal Avenues to Address the Plight of Victims of Climate Change: Problems and Prospects," *Journal of Environmental Law and Litigation* vol. 22 (2007) pp. 267–299; A. Williams, "Turning the Tide: Recognizing Climate Change Refugees in International Law," *Law and Policy* vol. 30 (2008) pp. 502–529; C. Voigt, "State Responsibility for Climate Change Damage," *Nordic Journal of International Law* vol. 77 (2008) pp. 1–22; and the Symposium on "Climate Change Liability and the Allocation of Risk," *Stanford Environmental Law Journal* vol. 26A (2007) pp. 3–334.
88. On the footprint metaphor (Chapter 1, note 12), see also W.E. Rees, "Ecological Footprint: Concept," in S.A. Levin (ed.), *Encyclopedia of Biodiversity* vol. 2 (San Diego/CA: California Academic Press 2001) pp. 229–244; M. Paterson and S. Dalby, "Empire's Ecological Tyreprints," *Environmental Politics* vol. 15 (2006) pp. 1–22; and Chapter 6 (note 4).

6 Epilogue: The Lords' Day?

This chapter subheading is borrowed from the thriller by Dobbs (Chapter 4, note 25), which also contains graphic descriptions of a Diego Garcia "rendition flight" and the detention of a suspected Islamist terrorist on the U.S. base (at pp. 111–112, 331–332, 341–342, 352–353, 362–363, 405–406).

1. House of Lords, Session 2007–2008 [2008] United Kingdom House of Lords Decisions 61, on an appeal brought by the UK government, from the court of appeal's decision of May 23, 2007 [2007], England and Wales Court of Appeal Reports (Civ.) 498 (see Chapter 2, note 62). See the summary, "Order Banning Chagos Islanders Not Unlawful," *Times* No. 69462 (London, October 23, 2008) p. 71; comments by Allen (Chapter 1, note 64), and P.H. Sand, case note in *American Journal of International Law* vol. 103 No. 2 (forthcoming 2009).
2. An Act to Remove Doubts as to the Validity of Colonial Laws, *28 & 29 Vict.* (1865), ch. 63; for background see D.B. Swinfen, *Imperial Control of Colonial Legislation 1813–1865: A Study of British Policy Towards Colonial Legislation Powers* (Oxford: Clarendon 1970).
3. The text of the concurring and dissenting opinions is not reproduced here and is available online at <http://www.publications.parliament.uk/pa/ld200708/ldjudgmt/jd081022/banc-10.htm>.
4. Credit for first use of the "imperial footprint" metaphor in connection with Diego Garcia is due to sociologist David Vine's thesis (Chapter 1, note 6). For historical analysis of the underlying "universalist ideology" and of American self-perception as "the advance guard of civilization, leading the way against backward and barbaric nations and empires," see R. Kagan, *Dangerous Nation* (New York: Knopf 2006) p. 416. The contemporary general literature on "Empire," however, has become too voluminous to be adequately addressed here; suffice it to note that more and more literature tends to leave emperors with less and less clothes. For example, see K. Roberts-Wray, *The Rise and Fall of the British Empire* (Nottingham: University of Nottingham Press 1969); R.W. Van Alstyne, *The Rising American Empire* (New York: Norton 1974); G. Vidal, *The Decline and Fall of the American Empire* (Berkeley/CA: Odonian Press 1992); M. Hardt and A. Negri, *Empire* (Cambridge/MA: Harvard University Press 2000); A.J. Bacevich, *American Empire: The Realities and Consequences of US Diplomacy* (Cambridge/MA: Harvard University Press 2002); E. Todd, *Après l'empire:*

Notes—Chapter 6

essai sur la décomposition du système américain (Paris: Gallimard 2002); H. Magdoff, *Imperialism Without Colonies* (New York: Monthly Review Press 2003); N. Ferguson, *Empire: The Rise and Demise of the British World Order and the Lessons for Global Power* (New York: Basic Books 2004); id., *Colossus: The Rise and Fall of the American Empire* (London: Allen Lane 2004); W.E. Odom and R. Dujarric, *America's Inadvertent Empire* (New Haven/CT: Yale University Press 2004); A. Bartholomew (ed.), *Empire's Law: The American Imperial Project and the 'War to Remake the World'* (London: Pluto Press 2006); C.P. David and D. Grondin (eds.), *Hegemony or Empire? The Redefinition of US Power under George W. Bush* (Aldershot: Ashgate 2006); and P. Brendon, *The Decline and Fall of the British Empire* (London: Cape 2007).

5. K. Roberts-Wray, *Commonwealth and Colonial Law* (London: Stevens 1966) p. 727, cited by Lord Rodger (para. 79). The "arcane" division into settled and conquered/ceded colonies has been labeled an "ancient and formal nicety" by Tomkins (Chapter 2, note 47) p. 579.
6. Lord Hoffmann (appendix X, p. **131**) para. 31, citing Halsbury's *Laws of England* vol. 6 (4th edn. reissue 2003) para. 823. Lord Bingham of Cornhill, in his dissenting opinion (para. 69), calls the royal prerogative to legislate by order-in-council an "anachronistic survival."
7. Lord Hoffmann, *ibid*. para. 35 (p. **132)**, rejecting as "extreme" the argument put forward by the UK government that the exercise of "prerogative legislation" is immune from judicial review, but suggesting judicial review "with a light touch" (para. 52 [p. **138**]); see also Lord Rodger of Earlsferry concurring, para. 105.
8. Lord Hoffmann, *ibid*. para. 58 (p. **140**); Lord Rodger of Earlsferry concurring, para. 109; and Lord Carswell concurring, para. 130.
9. Dissenting opinion, para. 72(3), referring to the U.S. State Department letters mentioned in Chapter 2 (notes 46, 53 and 61), which even in Lord Hoffmann's view "might be regarded as fanciful speculations" (para. 57). See also R.W. Cooper [LtCdr Royal Navy, ret.], letter to the editor, *Times* No. 68111 (London, June 25, 2004) p. 27: "Any objections to the occupation of the western islands arising from the present occupiers of Diego Garcia smack of military paranoia on the part of our allies, a phenomenon to which many of us have become accustomed. Defence security compromised? By whom—flying fish?"
10. See the Posford Haskoning study, Chapter 2 (note 60).
11. Lord Hoffmann *ibid*. para. 55 (p. **140** "funding is the subtext of what this case is about"); Lord Rodger of Earlsferry concurring, para. 113; and Lord Carswell concurring, para. 132. The resettlement cost forecasts of the FCO-commissioned studies have been challenged as exaggerated by the subsequent Howell Report (Chapter 2, note 60) pp. 31–33, whose alternative estimates of approximately £25 million ($37 million) over a period of 5 years were in turn criticized as too low by the Turner Report (Chapter 2, note 60) p. 6. See also Allen (Chapter 1, note 64) at pp. 687–689.
12. See the report of the Public Accounts Committee, Chapter 2 (note 66).
13. Chapter 29: "No freeman shall be taken, or imprisoned...or exiled, or any otherwise destroyed...but by the lawful judgment of his peers, or by the law of the land"; as quoted in Lord Hoffmann's opinion para. 42 (p. **135**), together with Blackstone's *Commentaries on the Laws of England* vol. 1 (15th edn. 1809) p. 137: "But no power on earth, except the authority of Parliament, can send any subject of England out of the land against his will; no, not even a criminal."
14. Lord Hoffmann *ibid*., para. 34 (p. **132**), and Lord Rodger of Earlsferry concurring, para. 86.
15. District of Columbia Court of Appeals, Chapter 2 (note 40).
16. Judge Janice Brown (D.C. Cir., April 21, 2006), Chapter 2 (note 42).
17. See Chapter 1 (note 67).
18. Lord Hoffmann, para. 64 (p. **142**); see also Chapter 2 (note 74).
19. See Chapter 1 (note 60); in contrast, the UK ratification of Geneva Convention Protocols I and II was formally extended to the BIOT, see Chapter 1 (note 61).

Notes—Chapter 6

20. See Chapter 1 (notes 62–64), and pp. **49**, **51**, and **58**. According to the FCO, 'the UK Freedom of Information Act has not been extended to the Overseas Territories' (letter to the author dated March 31, 2009, from BIOT Administrator J. Yeadon).
21. Lord Rodger of Earlsferry, concurring opinion (para. 85); see Chapter 1 (note 68).
22. See Chapter 2 (note 68).
23. Drower (Chapter 2, note 11) p. 65. On the ambivalent U.S. attitude toward decolonization efforts in the UN General Assembly, see Chapter 1 (note 15).
24. Lord Hoffmann, para. 49 (p. **137**); see also the concurring opinion of Lord Rodger of Earlsferry, para. 114; and *cf.* A. Twomey, "Responsible Government and the Divisibility of the Crown," *Public Law* [2008] pp. 742–767, at p. 766.
25. See Chapter 1 (note 14) and Chapter 2 (para. 5 of Exhibit 2, p. **20**, and p. **24**) on the efforts of the Foreign and Commonwealth Office to circumvent Article 73, now implicitly condoned by Lord Hoffmann in para. 55 of the majority opinion (p. **140**).
26. As one of his famous "fourteen points," which found its way into Article 22 of the League of Nations Covenant, 225 Consolidated Treaty Series 195, at p. 203. See N. Matz, "Civilization and the Mandate System under the League of Nations as Origins of Trusteeship," *Max Planck Yearbook of United Nations Law* vol. 9 (2005) pp. 47–96, at pp. 50, 71; A. Anghie, *Imperialism, Sovereignty and the Making of International Law* (Cambridge: Cambridge University Press 2005) p. 122; and M. Mazower, "An International Civilization? Empire, Internationalism and the Crisis of the Mid-Twentieth Century," *International Affairs* vol. 82 (2006) pp. 553–566, at p. 560.
27. M. Koskenniemi, *The Gentle Civilizer of Nations: The Rise and Fall of International Law 1870–1960* (Cambridge: Cambridge University Press 2002) p. 171.
28. See "Chagos Environment Network, *The Chagos Archipelago: Its Nature and the Future* (London: Chagos Conservation Trust 2009); *cf.* S. Gray, "Giant Marine Park Plan for Chagos: Islanders May Return To Be Environmental Wardens," *Independent* (London, online February 9, 2009); see also UK: 'Mega-MPA' Proposed for Indian Ocean Territory", *Environmental Policy and Law* vol. 39 (2009) pp. 75–76; "Ocean Blues," *Economist* (London, online February 9, 2009); and F. Pearce, "Conservations at the Expense of Homes?" *New Scientist* vol. 201 no. 2696 (February 21, 2009) p. 10. C.R.C. Sheppard and J. Turner, 'Eco-Imperialism', letter to the editor, *New Scientist* vol. 201 No. 2700 (March 21, 2009) p. 22.
29. Presidential Proclamations 8335–8337 of January 6, 2009, on the Marianas, Pacific Remote Atolls and Rose Atoll Marine National Monuments, 74 Federal Register 1555 (January 12, 2009). Most of the islands concerned have been administereded by the U.S. Navy or the Department of the Interior as colonial 'possessions' without self-government, with 'freedoms denied [to the indigenous Chamorro population] for the sake of empire'; R.D.K. Herman, 'Inscribing Empire: Guam and the War in the Pacific National Historical Park', *Political Geography* vol. 27 (2008) pp. 630–651, at p. 650; *cf.* R.F. Rogers, *Destiny's Landfall: A History of Guam* (Honolulu: University of Hawaii Press 1995). Unlike BIOT, however, these territories are at least included in the U.S. Government's periodic reports to the 'Committee of 24' of the United Nations under Article 73 of the Charter (see Chapter 1, note 14); see UN General Assembly Resolution 63/108 of December 18, 2008, part B/VI on Guam.
30. See the press statement by Olivier Bancoult, chairman of the Chagos Refugee Group, in Port Louis on February 13, 2002; *Agence de Presse Africaine* (SR/pm/APA, online February 13, 2002).
31. Possible structures of participation and self-government are outlined in the Howell report, Chapter 2 (note 60) pp. 29–30.
32. See pp. **15–16**, and Edis (Chapter 1, note 3) p. 39. See also Allen (Chapter 1, note 64) p. 693, on the application to BIOT of the 1966 International Convention on the Elimination of All Forms of Racial Discrimination (ICERD, 660 United Nations Treaty Series 195), in force January 1, 1969, ratified by the UK on March 7, 1969.

33. See the conventions listed at p. **64**.
34. See Chapter 1 (note 38). Cf. the official protest by Mauritian foreign minister Arvin Boolell, 'Mauritius Opposes Setting Up of a National Park in Chagos Archipelago', African Press Agency (APA on-line March 12, 2009), reaffirming the country's claim of sovereignty.
35. See the 1989 Agreement on Fishing in Mauritian Waters, quoted in Chapter 1 (note 47).
36. See the declarations in the UN General Assembly and in the British House of Commons, cited in Chapter 1 (note 40).
37. William Marsden CMG, Chagos Conservation Trust Chairman, former commissioner for BIOT and director of Americas, FCO, and former British ambassador to Argentina; as quoted by Gray (note 28).
38. See Chapter 1 (note 43).
39. The Indian Ocean Commission (IOC), established in 1984 with funding from the European Union (to the tune of 18 million Euros for the period 2006–2011), currently includes the Comoros, France (Réunion), Madagascar, Mauritius, and the Seychelles, with a "Regional Programme for the Sustainable Management of the Coastal Zones of the Indian Ocean" headquartered in Quatre Bornes, Mauritius (RECOMAP <http://www.recomap-io.org>); see K. Tetzlaff, "Indian Ocean Commission," *Yearbook of International Environmental Law* vol. 18 (2007) pp. 659–663, at p. 661.
40. See Chapter 3 (note 37).
41. See text in Chapter 3 (at note 39).
42. As indicated on the official BIOT website (FCO, last updated August 8, 2007).
43. See Gray (note 28), quoting Marsden (note 37).
44. See Chapter 1 (note 46).
45. See Chapter 1 (notes 50 and 56).
46. On the need to ensure compatibility of environmental restrictions with rights of passage under UNCLOS articles 58(1), 194(4), 211(5) and 220, see A. Merialdi, "Legal Restraints on Navigation in Marine Specially Protected Areas," in T. Scovazzi (ed.), *Marine Specially Protected Areas* (The Hague: Kluwer Law International 1999) pp. 29–34; and J.M. Van Dyke, "The Disappearing Right to Navigational Freedom in the Exclusive Economic Zone," *Marine Policy* vol. 29 (2005) pp. 107–121.
47. United Nations Convention on the Law of the Sea (1833 United Nations Treaty Series 397), Article 234.
48. See Sand, Chapter 1 (note 46) p. 375; and T. Scovazzi, "Recent Developments as Regards Maritime Delimitation in the Adriatic Sea," in R. Lagoni and D. Vignes (eds.), *Maritime Delimitation* (Leiden: Nijhoff 2006) pp. 189–203.
49. Designation of the Papahanaumokuakea "particularly sensitive sea area" (PSSA) by the IMO on April 3, 2008; see 73 Federal Register 73593 (December 3, 2008). Cf. J. Brax, "Zoning the Oceans: Using the National Marine Sanctuaries Act and the Antiquities Act to Establish Marine Reserves in America," *Ecology Law Quarterly* vol. 29 (2002) pp. 71–129; R. Lagoni, "Marine Protected Areas in the Exclusive Economic Zone," in A. Kirchner (ed.), *International Marine Environmental Law: Institutions, Implementation and Innovations* (The Hague: Kluwer Law International 2003) pp. 157–167; and R.K. Craig, "Are National Monuments Better Than National Marine Sanctuaries? U.S. Ocean Policy, Marine Protected Areas, and the Northwestern Hawaiian Islands," *Sustainable Development Law and Policy* vol. 7 (2006) pp. 27–31.
50. "Marine National Monuments," White House Press Release (January 6, 2009).
51. See B. Oxman, "The Territorial Temptation: A Siren Song at Sea," *American Journal of International Law* vol. 100 (2006) pp. 830–851. The reference is to the famous treatise by John Selden, *Mare Clausum seu de Dominio Maris* (London: Meighen 1635; English translation by M. Needham, *Of the Dominion, or, Ownership of the Sea*, London: DuGard 1652, reprint Clark/NJ: Lawbook Exchange 2004); see G.R. Russ and D.C. Zeller, "From *Mare Liberum* to *Mare Reservarum*," *Marine Policy* vol. 27 (2003) pp. 75–78;

M. Brito Vieira, "*Mare Liberum* vs. *Mare Clausum*: Grotius, Freitas, and Selden's Debate on Dominion Over the Seas," *Journal of the History of Ideas* vol. 64 (2003) pp. 361–377; and E. Franckx, "The 200-Mile Limit: Between Creeping Jurisdiction and Creeping Common Heritage," *Jahrbuch für Internationales Recht / German Yearbook of International Law* vol. 48 (2005) pp. 117–149.
52. See note 26. On the historical analogy between Wilson's "sacred trust of civilization" and trusteeship over coastal areas, see P.T. Fenn Jr., "Justinian and the Freedom of the Sea," *American Journal of International Law* vol. 19 (1925) pp. 716–727, at p. 724.
53. On U.S. environmental law, see the classic by J.L. Sax, "The Public Trust Doctrine in Natural Resources Law: Effective Judicial Intervention," *Michigan Law Review* vol. 68 (1970) pp. 471–556; for a comparative overview of the concept's spread to other legal systems worldwide, see P.H. Sand, "Sovereignty Bounded: Public Trusteeship for Common Pool Resources?," *Global Environmental Politics* vol. 4 (2004) pp. 47–71.
54. See M.C. Jarman, "The Public Trust Doctrine in the Exclusive Economic Zone," *Oregon Law Review* vol. 65 (1986) pp. 1–33; J.H. Archer and M.C. Jarman, "Sovereign Rights and Responsibilities: Applying Public Trust Principles for the Management of EEZ Space and Resources," *Ocean and Coastal Management* vol. 17 (1992) pp. 253–271; R.G. Hildreth, "The Public Trust Doctrine and Coastal and Ocean Resources Management," *Journal of Environmental Law and Litigation* vol. 8 (1993) pp. 221–236; and M. Turnipseed *et al.*, "The Silver Anniversary of the United States' Exclusive Economic Zone: Twenty-Five Years of Ocean Use and Abuse, and the Possibility of a Blue Water Public Trust Doctrine," *Ecology Law Quarterly* vol. 36 (2009) pp. 1–71 and M. Turnipseed *et al.*, 'Oceans: Legal Bedrock for Rebuilding America's Ecosystems', *Science* vol. 324 No. 5924 (April 10, 2009) pp. 183–184.
55. See P.H. Sand, "Public Trusteeship for the Oceans," in Ndiaye and Wolfrum (Chapter 1, note 59) pp. 521–543. *Cf.* the report of the Independent World Commission on the Oceans, *The Ocean: Our Future* (Cambridge: Cambridge University Press 1998) at pp. 45–46; the report of the Pew Oceans Commission, *America's Living Oceans: Charting a Course for Sea Change* (Philadelphia/PA: Pew Charitable Trust 2003) at p. 10; D.R. Christie, "Marine Reserves, the Public Trust Doctrine and Intergenerational Equity," *Journal of Land Use* vol. 19 (2004) pp. 427–434; and G. Osherenko, "New Discourses on Ocean Rights: Understanding Property Rights, the Public Trust, and Ocean Governance," *Journal of Environmental Law and Litigation* vol. 21 (2006) pp. 259–316.

Appendices

1. Concluded by exchange of notes signed at London on December 30, 1966, entered into force on December 30, 1966; text in 603 United Nations Treaty Series 273 (No. 8737, registered by the UK on August 22, 1967); 18 United States Treaties 28 (TIAS No. 6196); United Kingdom Treaty Series [1967] No. 15 (Cmnd. 3231); amended in 1976 (see appendix V, p **105**).
2. Copy certified by U.S. consul Ray E. White Jr. (at the American Embassy in London), on file with U.S. National Archives and Records Administration (Washington, D.C.), RG 59/150/64–65 (1964–1966, Box No. 1552), declassified on November, 16, 2005. On the question of the exact date, see Chapter 1 (note 27).
3. UK-U.S. Agreement concerning the United States Tracking and Telemetry Facilities in the Island of Mahe in the Seychelles, simultaneously concluded by exchange of notes (with Agreed Minute) also signed at London on December 30, 1966, by U.S. ambassador Bruce and Lord Chalfont, text in 603 United Nations Treaty Series 245 (No. 8736); superseded by the Agreement between the United States and the Seychelles Relating to a Tracking Station

on Mahe Island, signed at Victoria, Seychelles, on June 29, 1976, text in 1060 United Nations Treaty Series 96.
4. Agreed by exchange of notes signed at London on October 24, 1972, entered into force on October 24, 1972; text in 866 United Nations Treaty Series 302 (registered by the United Kingdom on April 11, 1973); 23 United States Treaties 3087 (TIAS No. 7481); United Kingdom Treaty Series [1972] No. 26 (Cmnd. 516); superseded by the 1976 Supplement, appendix IV (p. **91**).
5. 18 United States Treaties 575 (TIAS No. 6267), 193 United Nations Treaty Series 188; superseded in 1973 by the Malaga-Torremolinos Convention, 1209 United Nations Treaty Series 32.
6. Agreement concluded by exchange of notes signed at London on February, 25, 1976, entered into force on February 25, 1976; text in 1018 United Nations Treaty Series 372 (registered by the United Kingdom on July 27, 1976); 27 United States Treaties 315 (TIAS No. 8230); United Kingdom Treaty Series [1976] No. 19 (Cm. 6413).
7. Footnote reading: "The two Parties agree that the Supplementary Arrangements and the related notes are not international agreements, nor part of the Agreement of 25 February 1976, and it is therefore their view that they do not qualify for registration under article 102 of the Charter. They are published for information at the request of the Government of the United States of America."
8. Amendment agreed by exchange of notes signed at London on June 22 and 25, 1976, entered into force on June 28, 1976; text in 1032 United Nations Treaty Series 323 (registered by the United Kingdom on January 18, 1977); 27 United States Treaties 3448 (TIAS No. 8376); United Kingdom Treaty Series [1976] No. 88.
9. Concluded by exchange of notes signed at Washington, D.C., on December 13, 1982, entered into force on December 13, 1982; text in 2001 United Nations Treaty Series 397 (registered by the United States on January 20, 1998); 24 United States Treaties 4553 (TIAS No. 10616).
10. Concluded by exchange of notes signed at Washington, D.C. on November 16, 1987, entered into force on November 16, 1987; text in 1576 United Nations Treaty Series 179 (No. 27519, registered by the United Kingdom on August 24, 1990); United Kingdom Treaty Series [1988] No. 60.
11. Concluded by exchange of notes signed at London on June 18 and July 21, 1999, entered into force on July 21, 1999; text in 2106 United Nations Treaty Series 294 (No. 36672, registered by the United Kingdom on May 17, 2000); United Kingdom Treaty Series [2000] No. 1 (Cm. 4582).
12. The documents summarized in this appendix were released to the public by the UK Foreign and Commonwealth Office on September 2, 2008, pursuant to freedom-of-information request 0670–08.
13. House of Lords, Session 2007–2008, [2008] United Kingdom House of Lords Decisions 61, on an appeal brought by the UK government, from the court of appeal's decision of May 23, 2007, [2007] England and Wales Court of Appeal Reports (Civ.) 498 (see Chapter 2, note 62). The House of Lords' Appellate Committee, by a 3:2 majority judgment delivered on October 22, 2008, allowed the government's appeal. *Not* reproduced here are the two concurring opinions by Lord Rodger of Earlsferry and Lord Carswell and the two dissenting opinions by Lord Bingham of Cornhill and Lord Mance.

INDEX OF NAMES

Bold-face references indicate page numbers in the text and in the appendices; references separated by strokes (/) indicate Chapter/Note numbers.

Abbott, Chris, 5/79
Abram, Nerilie J., 5/72
Acland, Antony A., **110**
Adam, M. Shiham, 1/50
Ahlström, Christer, 3/46
Aid, Matthew M., 4/6
Akimoto, Kazumine, 1/59
Albinski, Henry S., 3/26
Albright, Madeleine K., **9**, 1/42
Aldrich, Richard J., 4/6, 4/8
Aldrich, Robert, 1/22
Allan, James N., **29**, 2/37
Allen, Philip M., 1/30
Allen, Stephen, 1/64, 2/45, 2/63, 2/73, 6/1, 6/11, 6/32
Allott, Philip, 3/4
Amos, Valerie A., **129**
Anand, J.P., 3/29, 3/41
Anctil, Robert, 4/16
Andersen, Walter K., 3/29, 3/41
Andrew, Christopher, 4/6
Angenot, Jean-Pierre, 2/4
Anghie, Antony, 6/26
Archer, Jack H., 6/54
Arkin, William M., 3/27, 3/57
Armstrong, Anne, **106**
Arny, Wayne, 5/84
Asquith, Julian, **16**, 2/59
Aust, Anthony I., **23**, 2/19, 2/73

Baba, Michiko, **ix**
Bacevich, Andrew J., 6/4
Bagley, David H., **102**, 5/33
Bailey, Kathleen, 3/46
Baker, James, 3/57
Baldwin, Gordon B., 1/68
Ball, Desmond, 4/7–8, 4/11
Ballance, Lisa T., 4/50
Balmond, Louis, 2/47
Bamford, James, 4/8–9, 4/14–15
Bancoult, Olivier, **29–30, 63–5, 122, 126, 129–30, 136–8, 140–1**, 2/7–9, 2/16, 2/38–40, 2/43, 2/53, 2/62–3, 2/73, 6/30
Bandjunis, Vytautas B., **36**, 1/6, 1/8, 1/10, 1/18, 1/27–8, 1/31, 1/35–6, 2/13, 2/17, 2/27, 2/30, 3/6–7, 3/9, 3/11, 3/16, 3/21, 3/25, 3/54, 4/7, 4/16, 5/13
Bangaroo, Sulina, 2/45
Barber, Stuart B., **1**, 1/6
Barkhuysen, Tom, 2/68
Barros, James, 1/15
Bartholomew, Amy, 6/4
Bateman, Sam, 1/59
Bates, Ed, 2/73
Beamish, Tony, 1/9
Beaumont, Peter, 3/43
Beck, Ulrich, 4/41

Bellamy, David J., **58**
Benson, Peter, 2/33
Berends, Helena, 5/6
Berger, Jon, 4/18
Bergquist, Amy, 4/22
Bernhardt, Rudolf, 1/64, 2/29, 2/57–8, 3/30, 3/35, 3/45, 3/48, 4/13
Beshoff, Pamela, 4/22
Besson, Samantha, 2/74
Bezboruah, Monoranjan, 1/6, 2/13, 2/26, 3/23, 3/26
Bhatt, Anita, 1/11, 2/13, 3/41
Bingham, Thomas H., **32**, **63**, **138**, 6/6
Birchard, Bruce, 3/40
Blackman, Donna K., 4/47
Blackstone, William, **135**
Blanton, Thomas, 4/42
Bloomfield, Lincoln P. Jr., **118**, 2/53
Boese, Wade, 3/46
Boolell, Arvin, 6/34
Bootle-Wilbraham, Roger, 3/25
Bose, Sugata, 3/3
Bouchard, Christian, 5/76
Bourne, Gilbert C., 5/2
Bowett, Derek W., 2/57
Bowman, J. Roger, 4/21
Bowman, Larry W., 3/7, 3/15, 3/20, 3/31, 4/8
Boyce, Michael C., 3/12
Boyd, Ian, 4/48
Bradley, Anthony W., 2/63
Brand, Kobie, 5/46
Braun, Dieter, 3/30
Brax, Jeff, 6/49
Brendon, Piers, 6/4
Bridgewater, Peter, 5/5
Brito Vieira, Monica, 6/51
Brooks, Gary L., 1/31
Brooks, Robert A., 2/57
Broucher, David, 1/72
Brown, Edward D., 2/58
Brown (George-Brown), George, **6**, **69**, 1/37
Brown, George S., **24**
Brown, Gordon, **9**, 4/37

Brown, Janice R., **29**, 2/42, 6/16
Brownlie, Ian, 2/58
Bruce, David K.E., **6**, **8**, **69**, **72–3**, **83**, 1/37
Bruce, Ian, 3/59
Bruch, Carl, **ix**
Bruha, Thomas, 3/48
Buergenthal, Thomas, 2/29
Burrell, R. Michael, 1/58, 3/24, 3/29
Burrows, William E., 4/11
Busby, Joshua W., 5/86
Bush, George W., **29**, **65**, 4/42, 6/4
Butler-Sloss, Elizabeth, **31**, **128**
Buzan, Barry, 3/30
Byers, Michael, 2/47
Byers, Roddick B., 3/30

Calabresi, Massimo, 5/23
Calder, Kent E., 1/58, 2/27, 3/2, 3/23, 3/58
Campbell, Duncan, 2/64–5, 4/36
Campbell, Kurt M., 5/86
Carey, Sean, 2/65, 2/76
Carr, Peter, 5/6
Carswell, Robert, **32**, **63**, 6/8, 6/11
Carter, James M., 3/8
Carter, Jimmy, 5/38
Carter, Peter, 1/77
Chalfont (Jones), Alun, **8**, **73**, **83**, 1/37
Charney, Jonathan I., 2/57
Chatterjee, Pratap, 3/8
Chaturvedi, Sanjay, 3/22
Chellapermal, A., 1/38
Chesterman, Simon, 4/7
Chew, Emrys, 3/3
Chomsky, Noam, **145**
Choper, Jesse H., 2/42
Chossudovsky, Michel, 4/20
Christakis, Théodore, 4/40
Christensen, April A., **ix**
Christie, Donna R., 6/55
Church, John A., 5/76
Churchill, George T., 2/26, 2/30, 2/40
Clark, Ian, 3/31
Clarke, Alan, 5/84

Index of Names

Clarke, Anthony P., **32**, **122**
Clementson, John, 3/20
Cloudview, H., 2/55
Cohen, Robin, 3/22
Connell, John, 1/22
Cook, Robin, **127**
Cooley, Alexander, 1/27
Cooper, R.W., 6/9
Corbin, Jeremy, 3/15
Cottrell, Alvin J., 1/58, 3/24, 3/29
Cox, Tara M., 4/49
Craig, Robin K., 6/49
Crawford, James, 1/39
Creasy, Darryl K., **ix**
Cresswell, Peter, **32**, **122**
Crook, John R., 1/70, 2/40, 2/42, 4/28
Cross, William, 4/54
Crow, Jonathan, **131–2**, **134**, **137–8**
Culshaw, Robert N., **118**, 2/53
Culver, John, **33**, 2/26–7, 3/24
Curtis, Mark, 2/23

Daggett, Stephen, 5/85
Dalby, Simon, 5/88
Dalyell, Tom, 1/9, 2/14
David, Charles-Philippe, 6/4
Davies, Chloe, **ix**
Davis, Peter, 4/18
De Boer, Marijke N., 4/53
De Grimaldo, Sayda, 5/54
De l'Estrac, Jean Claude, **29**, 1/20, 1/24, 1/39, 2/37
De Lupis, Ingrid D., 4/13
Denuzière, Maurice, 1/19
Deshpande, V.S., 3/31
De Silva Jayasuriya, Shihan, 2/4
Deslarzes, Kenneth J.P., 4/52
De Young, Cassandra, 1/50
De Zayas, Alfred M., 2/29
Diakov, Anatoli, 3/49
Di Mento, John M., 4/11, 4/49
Dipla, Maritini, 2/58
Dobbs, Michael, 4/25, 6/1
Domingues, Catia M., 5/72
Donovan, Gregory P., 4/53

Doughty, B.T., 3/10, 3/13, 3/19, 5/20, 5/22
Doward, Jamie, 4/38
Dowdy, William L., 1/58, 3/7, 3/26
Doyle, Timothy, 5/76
Doyon, Dennis F., 3/40
Drower, George M.F., 2/11, 6/23
Ducasse, Michel, **ix**, **145**
Dujarric, Robert, 6/4
Dussercle, Roger, 1/5
Dycus, Stephen, 4/42, 5/39

Edis, Richard, 1/3, 2/2, 2/24, 3/9, 3/18, 3/56–8, 4/3, 4/16–17, 4/52, 5/1, 5/14–15, 5/24–5, 5/49, 5/56, 6/32
Emery, Alan R., 5/2
Emery, Eleanor J., 2/14
Erickson, Andrew S., 1/11, 3/19, 3/29, 3/40, 3/47, 3/50, 3/59, 5/83, 5/85
Erickson, Larry E., 5/52
Evans, Rob, 2/47
Evers, Sandra J.T.M., 2/45

Fallon, Richard H., 2/43
Farish, William S., **116**
Farran, Sue, 2/62
Fawcett, James E., 2/18
Feldman, David, 2/74
Fenn, Percy T. Jr., 6/52
Ferguson, Niall, 6/4
Fieldhouse, Richard, 3/27
Fingar, Thomas, 5/86
Finnis, John, **134**, **137**
Firestone, Jeremy, 4/48
Fischer, David, 3/38, 3/45
Fischer, Joschka, 4/41
Fleming, L. Vin, 5/6, 5/60
Fletcher, Martin, 2/64
Forsberg, Steven J., **ix**, 2/4, 3/36, 4/2, 4/12
Fox, Hazel, 1/64, 2/21, 2/54, 2/59
Francioni, Francesco, 4/13
Franck, Thomas, 2/42
Franckx, Erik, 6/51

Index of Names

Freitas, Justo S. de, 6/51
Frosch, Robert A., 4/7, 5/78
Fuller, Jack, 5/23

George, Roger Z., 4/11
Gerson, Joseph, 3/40
Ghandi, Indira, **38**
Gibbs, Richard J.H., **30, 126**
Gifford, Richard, **ix**, 2/69, 2/74
Gillespie, Alexander, 5/87
Glennon, Michael J., 4/22
Goldblat, Jozef, 3/36
Goodman, Sherri, 5/78–80, 5/86
Goodwin-Gill, Guy, 2/18
Gore-Booth, Paul, **17**, 2/9, 2/59
Graham, Nick, 5/73
Grantham, Jack, **2**
Gray, Sadie, 4/50
Greatbatch, Bruce, **24**, 2/14
Greenhill, Denis, **17**
Greenwood, Anthony, **3**
Grenier, Tom, 5/16
Grey, Stephen, 4/35
Grojean, Charles D., 3/25
Grondin, David, 6/4
Grotius (de Groot), Hugo, 6/51
Guitart, Carlos, 5/55
Gupta, Joyeeta, 5/87

Hails, Sandra, **ix**
Hakimi, Monika, 4/28
Hall, Richard, 3/3
Halperin, Samuel, 4/6
Hansen, Jerry, 5/52–3
Hanson, Jeffrey A., 4/21
Harbury, Jennifer, 2/43
Hardt, Michael, 6/4
Harkavy, Robert E., 1/58, 3/2, 4/11, 4/16
Harrison, Kirby, 3/5
Harrison, Selig S., 1/24, 3/27, 3/29–30
Hart, Harry S., **2**
Harvey, Philip, 2/38
Harwood, Charles J. Jr., 1/51

Hatch, John, 2/76
Hattersley, Roy, **97, 104**
Haulman, Daniel L., 3/58
Hayden, Michael V., **48**, 2/43
Hayes, Peter, 3/26, 4/2
Henckaerts, Jean-Marie, 2/29
Henley, Jon, 5/76
Herman, R. Douglas K., 6/29
Hettling, Jana K., 4/17
Heyman, Philip B., 1/67
Heyning, John E., 4/50
Hildreth, Richard G., 6/54
Hill, Christopher, 1/9
Høegh-Guldberg, Ove, 5/73
Hoffmann, Leonard H., **32, 63–4, 122–43**, 2/73, 6/6–9, 6/11, 6/13–14, 6/18, 6/24–5
Holgate, Simon, 5/75
Holm, Kyrre, 3/44
Holmes, Harry A., **110**
Hooper, Anthony, **32, 122**, 2/53, 2/61
Hopkins, Michael F., 4/6
Houbert, Jean, 1/20, 3/22
Howe, Jonathan T., **108**
Howell, John, 2/60, 6/11, 6/31
Howells, Kim, 1/50
Hoyt, Edwin P., 1/3
Hubert, Kate, 2/55
Huckle, Alan E., **116–17**
Huff, Peter G., 1/3
Hughes, Simon, **ix**
Huls, Nick, 2/68
Hunt, Charles D. Jr., 5/50
Hunt, John, 3/26
Hunter, Andrea, 4/54
Hurd, Douglas R., **43**

Igel, Heiner, **ix**
Ingram, Adam, 1/71
Ioannou, Krateros, 3/45

Jacobs, Christopher W., 1/69
Jarman, M. Casey, 6/54
Jarvis, Christina, 4/48
Jasny, Michael, 4/49

Index of Names

Jawatkar, K.S., 1/27, 2/27, 3/24, 3/26, 3/29, 3/31, 4/3
Jeas, William C., 4/16
Jeffery, Laura, 2/33, 2/48, 5/77
Johannessen, Steffen F., 2/33
Johnson, Asa, **ix**
Johnson, Chalmers, 3/2, 3/18
Johnson, Robert D., 3/23
Johnson, Roy L., 1/6
Johnston, Barbara R., 2/38
Jonas, David S., 4/17
Jones, Rodney W., 3/41
Jones, Stephanie, 2/19
Justinian, I., 6/52

Kagan, Robert, 6/4
Kalugin, Oleg, 4/14
Kampbell, Donald H., 5/52–3
Kauffman, Tim, 5/84
Kaushik, Devendra, 3/30
Kay, David A., 1/15
Keckes, Stjepan, **ix**
Keefe, Patrick R., 4/7
Keller, Helen, 2/74
Kelley, Judith, 1/66
Kennedy, Edward M., **33**, 2/26, 3/23
Kennedy, John F., **5**, 5/78
Kentridge, Sydney, **131–2, 135–8**
Kershaw, Anthony, **89**
Khan, Jooneed, 3/22
Kirchner, Andree, 6/49
Kissinger, Henry, 3/26
Kline, David, 5/73
Kline, Robert D., 4/11
Kluge, Heinz, **ix**
Koetter, Wolfgang, 3/45
Koivurova, Timo, 5/87
Kokott, Juliane, 4/13
Koohi-Kamali, Farideh, **ix**
Koskenniemi, Martti, 6/27
Krock, Hans-Jürgen, 5/84
Kwiatkowska, Barbara, 2/58

Labs, Eric J., 1/76
Lader, Philip, **115**

Ladwig, Walter C., **ix**, 1/11
Lagoni, Rainer, 6/48–9
Laird, Melvin R., **29**
Lalonde, Suzanne, 1/25
Larmour, Edward N., **104**
Larus, Joel, 1/58, 3/28, 3/41
Lavalle, Roberto, 2/57
Laws, John G., **30, 126–7, 141**, 1/16, 2/7–9, 2/19–20, 2/22
Leatherwood, Stephen, 4/53
Le Clézio, Jean-Marie G., 2/24
Lefebvre, Jeffrey A., 3/7, 3/15, 3/20, 4/8
Lehner, Rick, 5/16
Leigh, Daniel P., 5/50, 5/53
Lemière, Paul, 2/10
Le Moyne, Irve C. Jr., **ix**, 5/48
Lennard, Donald E., 5/84
Levi, Yoel, 5/16
Levin, Simon A., 5/88
Lewis, Adrian, **145**
Leymarie, Philippe, 2/1
Lin, Karen, 2/40
Lindén, Olof, 5/74
Lindsay-Poland, John, 5/54
Lord, Carnes, 1/11, 1/58
Lowe, Sarah, 5/46
Lutz, Catherine A., 3/2, 5/77
Lygo, Raymond D., **102**, 5/33
Lynch, Timothy P., 1/58

MacAlister-Smith, Peter, 3/30, 3/33
MacAskill, Ewen, 2/47
Macmillan, Harold, **5**
Madeley, John, 2/31, 2/34–6
Magdoff, Harry, 6/4
Malaisé, Hakim, 2/47
Malloch-Brown, Mark, 4/37
Mance, Jonathan H., **32, 63**, 2/73
Mancham, James R., 1/11, 1/19, 1/52, 3/21
Mansfield, Mike, 3/23
Mansfield, William M., **131, 133**
Marsden, William, 6/37, 6/43
Marty, Dick, 4/28

Maslen, Stuart, 1/72, 1/74
Matadeen, Keshoe P., **29**
Matthews, Sue, 5/46
Matz, Nele, 6/26
Mayer, Jane, 1/67
Mazower, Mark, 6/26
McAllister, Lucy, **ix**
McArthur, Elizabeth, 4/22
McCaffrey, Barry, **47**
McCain, John S., **2**, 1/11
McCarthy, Elena, 4/48
McDermott, Patrick, 4/42
McKenna, Philip, 5/84
McNamara, Daniel T., 1/31
McNamara, Robert S., **29**, **64**, 1/35, 2/16, 2/38, 2/40, 2/43
Mensah, Thomas A., 1/59
Menzie, Charles, 5/13, 5/28, 5/30
Merialdi, Angelo, 6/46
Miasnikov, Eugene, 3/49
Mikolay, Justin D., 1/11
Milde, Michael, 4/35, 5/61
Miliband, David W., **48**, 4/26, 4/34, 4/38–9
Millar, Thomas B., 1/58
Mitchell, Austin, 2/66
Moor, Louise, 1/23, 2/74
Moorer, Thomas H., 3/24, 3/54
Morris, J. Richard S., 3/8
Morris, Ted A., **ix**, 3/18, 3/55, 3/58, 3/60, 5/16, 5/50–1
Moules, Richard, 2/63
Mueller, Harald, 3/45
Munn, Margaret, **48**

Nagchoudhary, B.D., 4/3
Nasheed, Mohamed, 5/76
Nauvel, Christian, 2/38
Ndiaye, Tafsir M., 1/59, 6/55
Needham, Marchamont, 6/51
Negri, Antonio, 6/4
Nelson, Jay, **ix**
Neuberger, David E., **31**, **128**
Newsom, Eric, 2/40, 2/46, 3/56
Nicholls, Donald J., **137**

Nitze, Paul, 2/15
Nixon, Richard M., 1/35
Norton-Taylor, Richard, 4/20, 4/36
Nowak, Manfred, 4/22

O'Brien, Mike, 1/52
Odom, William E., 6/4
O'Hanlon, Michael E., 3/2
Okal, Emile A., 4/21
O'Keefe, Roger, 2/48
Ollivry, Thierry, 1/39, 3/35, 5/5
Oraison, André, 1/13, 1/18, 1/20, 2/31, 2/36, 3/19
O'Rourke, Ronald, 3/20
Orr, Iain, 5/5
O'Shea, Michael, 1/50
Osherenko, Gail, 6/55
Osofsky, Hari M., 5/87
Ottaway, David B., 2/28
Ouseley, Duncan B.W., **31**, **124**, **127**, **129**, **140**
Oxman, Bernard, 6/51

Pack, Bradley, 4/42
Pakenham, Frank, **16**
Pallawatta, Nirmali, 5/46
Palmer, Norman D., 3/29
Palmer, Stephanie, 2/47, 2/76
Panikhar, S.K.M., 3/3
Paquin, Joan, 5/57
Parris, Matthew, 2/14
Parry, Clive, 1/2
Parsons, E. Chris M., 4/49
Patel, Shenaz, 2/33, 3/15
Paterson, Matthew, 5/88
Peake, Hayden B., 4/6
Pearce, Fred, 5/7, 5/68, 5/70, 6/28
Pears, Jonathan, 5/6
Pell, Claiborne, 3/23
Pelzer, Norbert, 3/35
Pepi, Vanessa E., 5/41
Perreau-Saussine, Amanda, 2/76
Perrin, William F., 4/50
Petit, Jérôme, 5/74
Pfeffer, W. Tad, 5/75

Index of Names

Pforzheimer, Walter, 4/6
Pienkowski, Michael W., 5/10, 5/68, 5/82
Pilger, John, 1/7, 1/17, 2/2, 2/15, 2/19, 2/25, 4/20
Pillai, Priva, 1/74
Pinto, M.C.W., 3/30
Plender, Richard, 2/18
Pope, Frank, 5/2, 5/46
Powell, Geoffrey, 2/4
Price, Andrew R.G., **ix**
Pridham, Charles, 2/5
Proctor, Dena, 5/6, 5/60
Proechel, Tania M., 5/54
Prosser, Albert Russel G., **25**, 2/33

Quinn, Terrence M., 5/50

Rahm, Richard, **ix**
Rahmstorf, Stefan, 5/75
Rais, Rasul B., 1/35, 3/23, 5/22
Ramesh, Randeep, 5/76
Ramgoolam, Navinchandra, **9**, 4/37
Ramgoolam, Seewoosagur, 1/20
Rammell, Bill, **33**, **129**, **140**, 2/52
Raven, John, 5/73
Rawcliffe, John, 5/40
Rees, William E., 5/88
Reid, Karen, 2/68
Reynolds, Joel R., 4/48
Rice, Condoleezza, **48**
Rich, Monty, 4/9
Richelson, Jeffrey T., 4/7–8, 4/10–11, 4/16
Ridenour, Andrew, 2/38
Riebesell, Ulf, 5/73
Rigo Sureda, Andrés, 1/30
Risely, Melissa, 5/76
Rivero, Horacio Jr., **2**, 1/10
Rivers, Julie, 5/46
Roberts, Alasdair, 4/41
Roberts, Bob, 4/38
Roberts, Colin, 1/31
Roberts-Wray, Kenneth O., 6/4–5
Robinson, Nicholas A., 4/42

Rodger, Alan, **32**, **63–4**, 6/5, 6/8, 6/11, 6/14, 6/21, 6/24
Rose, Stephen A., 1/59
Rosen, Mark E., 3/38, 3/57
Rouse, Joseph H., 1/68
Rousseau, Charles, 1/22, 1/27
Rubin, Alfred P., 4/13
Rudney, Robert, 3/46
Rumley, Dennis, 3/22
Rumsfeld, Donald H., **29**, 3/2
Russ, Garry R., 6/51

Saddul, Prem, **145**
Sand, Peter H., 1/43, 1/46, 2/69, 4/43, 6/53, 6/55
Sandars, Christopher T., 1/9, 1/35
Sands, Philippe J., 1/67
Santos Vara, Juan, 4/22
Satake, Kenji, 4/21
Sax, Joseph L., 6/53
Scharpf, Fritz W., 2/42
Schiffer, Lois, 5/39
Schlesinger, James R., **29**
Schonberg, Karl K., 3/38
Schubert, Renate, 5/86
Schwartz, Stephen I., 3/41
Scott, David A., 2/16
Scott, Karen N., 4/48
Scott, Robert, 1/2, 2/2
Scovazzi, Tullio, 6/46, 6/48
Scutro, Andrei, 3/49
Seaward, Mark R.D., 5/2
Sedley, Stephen J., **31–2**, **122**, **128**, **131**, **139**
Selden, John, **67**, 6/51
Shaker, Mohamed I., 3/45
Sheppard, Charles R.C., **ix**, 1/46, 3/17, 4/19, 5/2, 5/7, 5/29–31, 5/37, 5/46, 5/48, 5/55, 5/59, 5/72–4, 6/28
Sheridan, Bernard, 2/35, 2/68
Sherman, Robert B., 2/33
Sick, Gary S., 2/26
Simmons, Adele S., 1/20
Simpson, A.W. Brian, 1/23, 2/18, 2/69, 2/73–4, 2/76

Index of Names

Siophe, Hélène, 2/32
Slessor, Tim, 2/14, 2/22, 2/35
Slyomovics, Susan, 2/38
Snowe, Olympia S., 4/20
Snoxell, David R., **ix**, 2/15, 2/76, 3/61
Sohm, Earl D., **90**
Sohn, Louis B., 2/29
Sonnett, Neal R., 1/67
Soons, Alfred H.A., 2/58
Spalding, Mark, 1/46, 1/48, 5/2, 5/32, 5/37, 5/48, 5/59
Sparrow, Andrew, 4/38
Spiers, Ronald I., **98**, **104**, 3/1
Stafford Smith, Clive A., **ix**, 4/27, 4/36
Stein, Ted L., 3/56
Steyn, Johan, 1/67
Stoddart, David R., 1/9, 1/12, 5/2
Stoddart, Jonathan D., 2/15
Stone, Carolyn J., 4/48
Stummel, Dieter, 2/58
Subrahmanyam, K., 1/24, 3/27, 3/29–30
Sumida, Karen, **ix**
Sur, Serge, 4/22
Sweet, Alec S., 2/74
Swinfen, David B., 6/2
Sylva, Hervé, 2/33
Symons, Elizabeth, 5/35, 5/61

Taylor, Barbara L., 4/50
Taylor, Donald, 2/3, 5/10
Taylor, John D., 1/12, 5/2
Tetzlaff, Kerry, 6/39
Tigar, Michael E., 2/67
Todd, Emmanuel, 6/4
Tomkins, Adam, 2/47, 6/5
Topp, John M.W., 5/1–2, 5/37, 5/58
Toussaint, Auguste, 1/4
Towle, Philip, 3/31
Tribble, Gordon W., 5/52
Trinnaman, Judy, 5/84
Trood, Russell B., 1/58, 3/7, 3/26
Tucker, T., 3/10, 3/13, 4/5, 5/20, 5/22
Turner, John R., 2/60, 5/74, 6/11, 6/28
Turnipseed, Mary, 6/54

Twomey, Anne, 6/24
Tyrie, Andrew, 4/36

Umozurike, U. Oji, 1/14
Urish, D.W., 3/5
Urquhart, Conal, 3/43

Vacher, H. Leonard, 5/50
Valencia, Mark J., 1/59
Váli, Ferenc A., 2/27
Van Alstyne, Richard W., 6/4
Van der Wilt, Harmen G., 1/66
Van Dyke, Jon M., 1/59, 2/57, 3/4, 4/48, 6/46
Van Emmerik, Michiel, 2/68
Vencatessen, Michael, **125**, **128**
Verheyen, Roda, 5/87
Verkaik, Robert, 4/40
Vidal, Gore, 6/4
Vignes, Daniel, 6/48
Vine, David, **ix**, 1/6, 1/10, 2/3, 2/15, 2/25, 2/33, 2/38, 2/65, 2/76, 3/15, 5/77, 6/4
Voigt, Christina, 5/87

Walker, Iain B., 1/12, 2/32
Walker, John (aka Harper, James), 4/15
Waller, George M., **32**, **122**, **131**
Watt, Ian, **18**, **23**, 2/16
Weiss, Seymour, 2/27, 3/25, 3/54
Weissbrodt, David, 4/22
Wenban-Smith, Nigel, **108**
Westra, Laura, 2/38
Whistler, W. Arthur, 5/46
White, Charles J.B., **115**
White, Ray E., 1/27
Whitman, Bryan, 3/50
Whitman, Edward C., 4/11
Whitworth, Jerry A., 4/15
Widome, Daniel, 3/2
Wiebe, Virgil, 1/74
Wiebes, Cees, 4/6
Wilde, Ralph, 1/67
Wilkinson, Richard, 2/46, 3/56
Williams, Angela, 5/87

Williams, James, 5/69
Wilson, Charles, **117**
Wilson, Woodrow, **65**, 6/52
Winchester, Simon, 1/12, 1/16, 2/25, 2/77, 3/2, 3/21
Winterbottom, Rick, 5/2
Wolfrum, Rüdiger, 1/59, 6/55
Woodliffe, John, 1/27, 1/68
Wooldridge, Frank, 1/25
Wright, Jerauld, **2**
Wright, Patrick R.H., 2/9

Wu, Chih, 5/84

Yamaguchi, Mari, 3/52
Yeadon, Joanne, **ix**, 1/31, 4/45, 6/20

Zagorin, Adam, 4/38
Zeller, Dirk C., 6/51
Zorgbibe, Charles, 2/32
Zuckert, Eugene, 1/35
Zumwalt, Elmo R., 2/15, 2/27, 3/24, 3/39

SUBJECT INDEX

Bold-face references indicate page numbers in the text and in the appendices; references separated by strokes (/) indicate Chapter/Note numbers.

Aarhus Convention, *see* conventions
abandonment deeds, **25–7**, **125**, 2/34–5
abuse of power, **131**, **140**
Adriatic Sea, 6/48
Aero Contractors Ltd., 4/35
Afghanistan, **41**, **67**
African nuclear-weapon-free-zone, *see* conventions
Agalega Island, **18**, **21–2**, **122**, 1/4, 1/33
airfield Diego Garcia, **37**, **40–1**, **56**, **71**, **82**, **84–5**, **87**, **91**, **94**, **103**, **117–18**, 2, 5
Aldabra Island, 2, **5**, **22**, **69**, **105**, 1/9, 1/19, 2/11
Alien Tort Claims Act (USA), **29**, 2/38
American Petroleum Institute (API), 5/57
ammunition, **11–12**, **37**, **52**, **103**, **116–17**, 1/69–76
Amnesty International (NGO), 4/33
Andaman earthquake, **45**, 4/19–21
Antarctica, *see* conventions
anti-personnel mines, *see* conventions
Arctic claims, **9**
Argentina, 3/37, 6/37
Arms Control and Disarmament Agency (USA), **39**, 3/38
Ascension Island, 5/69
Atkins (W.S.) PLC, 3/12

Australia, **37**, **44**, **138**, 1/69, 3/26, 4/7, 5/76

Bahamas, **49**
'belongers', **17–18**, **23**, 2/19
Bermuda, **59**
Big Iron Trading Co., 3/13
biodiversity (biological diversity), **51**, **58**, **64**, **66**, 5/2, 5/6, 5/8, 5/69, 5/88
bioremediation, 'bioslurping', **56**, 5/52–3
'black hole' doctrines, **10–11**, **33**, **39**, **52**, **55**, **59**, **64**, 1/48, 1/67, 2/69, 3/36, 3/49
Boddan Island, 2/55
bombers, **4**, **37**, **40–1**
British Bar Human Rights Committee (BHRC), **13**, 1/77
British Virgin Islands, 5/69

California, **49**, **52**
Cameco Corp., 3/50
Canada, **40**, **44**, **67**, 1/69, 3/50, 4/7
Canary Islands, **49**
carbon dioxide (CO_2) emissions, **59**, **61**, 5/73
carrier force, **37**, 1/11, 3/18
Cayman Islands, 5/69

Center for Naval Analyses (CNA – USA), **59**, 5/78
Central Intelligence Agency (CIA – USA), **9**, **43**, **48**, 1/51, 4/16, 4/34–5, 5/61
cetaceans, *see* whales
Ceylon, *see* Sri Lanka
Chagos Conservation Management Plan (BIOT), **40**, **57**, 4/53, 5/59
Chagos Conservation Trust (CCT), **56**, **66**, 2/60, 5/2, 5/5, 5/50, 5/72, 5/74, 6/28, 6/37
Chagos Environment Network (NGO), **65**, 6/28
Chagossians, **1**, **15–33**, **37**, **63–6**, **123–7**, **129**, **131**, **137–42**, 2/15, 2/24, 2/29, 2/32–3, 2/38, 2/44, 2/48, 2/64, 2/68, 2/76, 5/77, 6/1
chemical weapons, *see* conventions
China, **1**, 1/11, 3/29, 4/43
citizenship, **20**, **23**, **29–30**, **123**, 2/44
civilian work force in Diego Garcia, **35–7**, **45**, **53**, 3/16
Claims Agreement UK/Mauritius, **25–9**, **125**, **128**, 2/37
'Classic Wizard', **44**
climate change, **59**, **128**, 5/72–9, 5/86–7
cluster bombs, *see* conventions
coconut crab, **52**, 5/12
Code of Conduct on Arms Exports (European Union), **39**, 3/44
cold war, **1**, **38**, **44**, **49**, 3/23, 4/6–7
Collins & Co., 2/46
colonialism, **2**, **63**, **65–6**, 1/22–3, 1/25, 2/5, 2/47, 2/76, 6/2, 6/5, 6/28
Colonial Laws Validity Act (UK), **63**, **132–4**, 2/64, 6/2
Commanding Officer Diego Garcia (USA), **39**, **54–5**, **79**, **86**, **89**, **92**, **95–6**, 1/31
Commissioner for BIOT, **4**, **16**, **24**, **70**, **86–7**, **89**, **92–3**, **95–6**, **123**, 1/31, 3/61, 6/37

'Committee of Twenty-Four' (UN), **2**, **17**, 1/14, 1/30, 6/29
Comoros, 6/39
compensation, **25–30**, **79**, **125**, **127–9**, **139**
Congress (USA), **5**, **18**, **20**, **24–5**, **35–6**, **40**, **43**, **88**, **96**, **124**, 1/32, 2/26–7, 3/1, 3/23–5, 4/5, 5/84, 5/86
conservation policy, **51**, **86**, **94**, **107**
contractors, **27**, **35–6**, **40**, **45–6**, **71–5**, **83**, **85**, **87**, **95**, **107–10**, 3/8, 3/11–14, 4/12, 5/33, 5/44, 5/51
conventions (multilateral agreements, treaties)
  access to environmental information (Aarhus), **49**, **64**, **66**, 4/44–5
  African nuclear-weapon-free zone (Pelindaba), **38–9**, 3/34–6
  Antarctic, **39**, **66**, 1/31, 3/37
  anti-personnel landmines (Ottawa), **11**, **13**, 1/69–70, 1/72, 1/74, 1/76
  biological diversity (CBD), **51**, **58**, **64**, **66**, 1/45, 5/8, 5/69
  chemical weapons, **11**, 1/45, 1/73
  cluster bombs/munitions (Dublin), **11**, 1/74, 5/82
  crimes against internationally protected persons, 1/45
  cultural property, 1/43
  Eastern African marine environment (Nairobi), **9**, **66**, 1/43
  endangered species (CITES), 1/45
  European convention against torture, **33**, **64**, 1/63
  European Convention on Human Rights (ECHR), **23**, **33**, **64**, **142**, 1/23, 2/18, 2/73
  Geneva conventions on humanitarian law, **10**, **33**, **64**, 1/60–1, 6/19
  human rights covenants, **10**, **33**, **64**, **66**, 1/62, 1/64, 2/18, 2/73
  law of the sea (UNCLOS), **9–10**, **32**, **58**, **67**, 1/46–9, 2/57–8, 6/46–7

Subject Index

migratory species (CMS), **49**, 1/45, 4/51
nuclear non-proliferation, **40**, 3/45
nuclear test ban (CTBT), **45**, **111–14**, 4/17
ozone layer, **9**, 1/45
racial discrimination (ICERD), 6/32
Southern Indian Ocean Fisheries (SIOFA), **9**, 1/48
torture (CAT), **10**, **33**, **46**, **64**, 1/63, 4/22
transfer of sentenced persons, **9**, 1/44
wetlands (Ramsar), **51–2**, **58**, 5/9–10, 5/21, 5/65–6, 5/82
whaling, **49**, 4/53
world cultural and natural heritage (UNESCO), **51**, **67**, 5/4–5
copra plantations, **1**, **3**, **18**, **21**, **27**, **122–3**, **128**, 1/4, 2/12, 2/15
'coral harvesting', **52–5**, 5/44
coral mortality, **59**, 5/73–4
coral reef, **2**, **52–5**, **59**, **108**, 3/17, 4/19, 5/1–2, 5/35, 5/74
Council of Europe, **47**, 2/74, 4/22, 4/28
Croatia, **67**
cryptology, **44**, 4/9

Day & Zimmermann Inc., 3/12
De Chazal du Mée (DCDM), 2/39
decolonization, **2**, **4**, **63**, **65**, **67**, 1/15, 1/20, 1/30, 2/65, 6/23
Denmark, 4/36
de-nuclearization, **37–8**, 3/28, 3/34–6, 3/41
Department of Defense (DoD – USA), *see* Pentagon
Department of Environment, Food and Rural Affairs (DEFRA – UK), 4/45, 5/66, 5/69
Department of State (USA), **9**, **18**, **25**, **30–2**, **40**, **48–9**, **66**, **107**, **109–10**, **117–18**, 1/35, 1/42, 2/15, 2/26, 2/46, 2/53, 3/1, 3/26, 3/54, 4/1, 4/20, 4/43, 5/83, 6/9

de-population/deportation, *see* expulsion
desalination, 5/84
Desroches Island, **3**, **5**, **69**, **105**
'detachment costs', **5–6**, **8**, 1/33
detention facility, **46–8**, **133**, 4/25–6, 4/36, 6/1
  see also prison ships
DG21 LLC, 3/12, 3/14
DGM21 LLC, **40**
Diana Princess of Wales Memorial Fund, 1/71
dredging, **52–5**, **58**, **108**, 5/21, 5/28, 5/35, 5/44
'dualist' doctrine in international treaty law, **143**, 2/73
Dublin Convention, *see* conventions
dynamiting, *see* reef-blasting

Echelon, **44**, 4/7
eco-imperialism, 6/28
EG&G Environmental Consultants, 5/28
El Niño, **59**
Emden, **1**, 1/3
'empire', **1**, **35**, **61**, **63**, **65**, 1/6–7, 1/9, 2/8, 3/2–3, 5/68, 5/88, 6/4, 6/26, 6/28
environmental damage, **52–6**, **108**, 4/19, 5/35, 5/56, 5/66, 5/87
environmental impact assessment (EIA), **49**, **53–5**, **108**, **113**, 4/54, 5/39, 5/43–4, 5/59, 5/84
Environmental Protection Agency (EPA – USA), **55**
environment charter (BIOT), 5/68
environment protection zone (BIOT), **9**, **67**, 1/46
Eritrea. 4/8
espionage, **44**, **76**, 4/11, 4/13, 4/15
Ethiopia, 4/8
European Convention on Human Rights, *see* conventions
European Court of Human Rights, **33**, **65**, 2/68

European Union (EU), **9**, **29**, **39**, **66**, 1/47–8, 3/44, 5/6, 5/74, 6/39
'excision', **3–4**, **8**
see also detachment costs
exclusive economic zone (EEZ), **9–10**, **32**, **66–7**, 1/46–8, 1/50, 1/56, 1/59, 2/58, 6/46, 6/49, 6/54
explosive safety quantity distance (ESQD) zone, **11–12**, **116–17**, 1/75
expulsion, **17–25**, **30–1**, **33**, **126**, **128**, 2/18, 2/29

Falkland Islands, 2/31, 5/69
Farquhar Island, **3**, **5**, **69**, **105**
feasibility studies, **32**, **59**, **63**, **126**, **128**, **130**, **141**, 2/60, 6/10–11
FEBROE Co., 5/51
First Support Services Inc., 3/12
fisheries, **9**, **49**, **87**, **95**, **102**, **123**, **128**, 1/46–50
Food and Agriculture Organization of the United Nations (FAO), 1/48, 1/50
'footprint of freedom', **2**, **61**, 1/12, 1/77, 5/88, 6/4
Foreign and Commonwealth Office (FCO – UK), **5**, **18**, **23**, **30–3**, **43**, **45**, **48–9**, **51**, **57–8**, **63**, **66–7**, **84**, **90–1**, **98**, **102**, **104–5**, **111**, **115–18**, **122**, **126–7**, **129–31**, **140–1**, **191**, 1/28, 1/33, 1/35, 1/37, 1/64, 1/71, 1/77, 2/14, 2/16, 2/37, 2/44, 2/46, 2/51–3, 2/60, 2/62, 2/66, 2/73, 3/35, 4/29–34, 4/37, 4/39, 4/45, 5/1, 5/6, 5/8, 5/35, 5/62–3, 5/68, 6/25, 6/37, 6/42
France, **1**, **15**, **38**, **67**, **122**, 1/2, 1/47, 6/39
Freedom of Information Act (FOIA – USA, UK), **41**, **48–9**, **121**, **191**, 1/75, 4/40–2, 5/26–8, 5/43–4, 5/48, 5/84, 6/20
fuel spills, **56**, **58**, 5/51–4

General Accounting Office (GAO – USA), **5**, **35**, **46**, 1/33, 1/36, 1/69, 3/11, 4/23, 5/19
Geneva Conventions, *see* conventions
Germany, **1**, **39**, 1/74, 3/44, 5/86
Ghyben-Herzberg freshwater lenses, 5/50
Gibraltar, 1/63, 4/45
'global posture review', **59**
Greece, 4/35, 5/51
'green enclosure', **67**, 1/46
Ground-Based Electro-Optical Deep Surveillance System (GEODSS), **12**, **44**, 4/16
Group Refizié Chagos (Chagos Refugees Group – Mauritius), **30**, 6/30
Guam, 3/52, 6/29
Guantánamo Bay (Cuba), **10**, **46**, **64**, 1/67, 2/69, 4/27
Guernsey, 1/63
Gulf Wars, *see* Middle East conflicts

Halliburton Corp., **29**, **35**, **46**, 2/40, 3/8
Hawaii, **67**, 5/84, 6/49
'hiding-hand' strategy, **49**
House of Commons (UK), **32**, **48**, **130**, **140**, 1/9, 1/28, 1/33, 1/40, 1/47, 1/64, 1/71, 2/14, 2/44, 2/51–2, 2/74–5, 3/15, 4/17, 4/25–6, 4/31–2, 4/34, 4/37, 4/39, 5/2, 5/62–3, 6/36
House of Lords (UK), **32**, **63–4**, **133**, **191**, 1/46, 2/44, 2/76, 3/25, 4/29, 4/33, 5/35, 6/1
human rights, **10**, **13**, **14**, **33**, **64**, **66**, **132**, **135–6**, **138**, **142–3**, 1/23, 1/62, 1/64, 1/67, 2/62, 2/69, 2/74–6, 4/25
see also conventions
hydro-acoustic monitoring, **45**, **49**, **111**, 4/17, 4/21

Îlois, *see* Chagossians
Immigration Orders/Ordinances (BIOT), **23–4**, **30–1**, **63–4**, **124–7**, **130**, **139**, **141**

Subject Index

Incorporated Research Institutions for Seismology (IRIS), **45–6**, 4/18, 4/21
Independent World Commission on the Oceans (IWCO), 6/55
India, **26**, **38**, **122**, 1/11, 1/24, 3/3, 3/27, 3/29, 5/76
Indian Ocean Commission (IOC), **66**, 6/39
Indian Ocean Sanctuary of the IWC, **49**, 4/53
information disclosure, **41**, **43**, **48–9**, **58**, **127**, 1/37, 1/75, 3/10, 3/60–1, 4/41–3, 5/26–8, 5/41, 5/43–4, 5/48
see also Freedom of Information Act
International Campaign to Ban Landmines (ICBL), 1/69–71, 1/77
International Committee of the Red Cross (ICRC), 1/72
International Court of Justice (ICJ), **4**, 3/53, 5/66
International Covenants on Human Rights, *see* conventions
International Criminal Court, **10**, **64**, 1/65–6
International Maritime Organization (IMO), **67**, 6/49
International Telecommunications Union (ITU), **86**, **94**
International Union for Conservation of Nature (IUCN), *see* World Conservation Union
International Whaling Commission (IWC), **49**, 4/53
'interoperability', 1/69, 1/74
invasive species, **55**, **58**, 5/46, 5/69
Iran, **41**, 3/56, 3/59
Iraq, **40–1**, **56**, **59**, **145**, 3/8, 3/50
Israel, **39**, 3/43, 3/46, 3/50
Italy, **67**, 3/47

Japan, **1**, **35**, **40**, **49**, **53**, 1/47, 3/7, 3/23, 5/12, 5/44
Jesuits, *see* missionaries
Joint Nature Conservation Committee (UK), **51**, 4/48, 5/6

joint ventures (UK–US), **35–6**, **109**, 3/12, 3/14, 5/84
judicial review, *see* justiciability
jurisdiction, **10–11**, **57**, **67**, **83**, 1/68, 1/74
justiciability, **29**, **63–4**, **126**, **134–5**, **138**, **141**, 2/42–3, 6/7
see also 'prerogative rights'

Kagnew station (Eritrea), **44**, 4/8
Kellogg, Brown & Root (KBR), **46**, 3/8
KGB (USSR), **44**, 4/14

Labour Party (UK), **37**, 3/28
LALIT [La Lutte] (NGO – Mauritius), **129**, 3/16
land deals, **3**, **6**, **8**, **18**, **22**, **81**, **88**, **96**, **124**, 1/19, 2/10
landmines, **11–13**, 1/69–70, 1/72, 1/74–6, 5/82
see also conventions
land use map Diego Garcia, **12**, 1/75, 3/60, 5/82
law of the sea, *see* conventions
League of Nations, 6/26
legal costs, **32**, **64**
Leucaena leucocephala (tangan-tangan), **55**, 5/46
Limuria, 1/2–3
Lockheed Martin Corp., 3/12

Madagascar, 6/39
Magna Carta, **64**, **135**, 2/47
Mahé airport (Seychelles), **3**, **22**, 1/33
Malaysia, **53**, **55**
Maldives, **9**, **32**, **67**, 1/50, 4/50, 5/76
Malta, **2**
mare nostrum, **35**, 3/4
Mariana islands, 6/29
marine mammals, **49**, 4/48–51, 4/54
marine park/reserve, **66–7**, 6/28, 6/34, 6/46, 6/49–50
massive ordnance penetrators (MOP), **41**

Subject Index

Mauritius, **x**, **2–5**, **8–10**, **15–17**, **19–29**, **33**, **37–8**, **58**, **66–7**, **71**, **87**, **122–3**, **125**, **142**, **145**, 1/10, 1/18, 1/20, 1/26, 1/33, 1/38–9, 1/41–5, 1/47–9, 1/51, 1/56–7, 1/60–1, 1/63, 1/65, 1/70, 1/73, 2/5, 2/10, 2/24, 2/31, 2/33, 2/37, 2/39, 2/65, 2/74, 3/16, 4/51, 5/4, 5/9, 6/34–5, 6/39

Mauritius Marine Conservation Society (MMCS), 5/5

Middle East conflicts, **35**, **39**, 1/69, 3/57

military bases, **ix**, **4**, **10**, **13**, **30**, **35–41**, **43**, **46**, **49**, **51–3**, **56**, **58–61**, **122**, **140**, 1/6, 1/9, 1/27, 2/26, 3/2, 3/24–5, 3/40–1, 3/47, 5/54

military construction, **18**, **23**, **35**, **39**, **41**, **52**, **55**, **58–9**, **81**, **85**, **91–2**, **103–4**, **109**, **111–13**, **116–18**, 3/7, 3/11, 3/47, 3/44, 5/84–5

military security, **18–19**, **30–1**, **41**, **43–5**, **48**, **59–60**, **63**, **76**, **79**, **84**, **86**, **92–4**, **120**, **130**, **140**, 1/37, 2/61, 4/6–7, 4/11–12, 4/40, 5/78–9, 5/86, 6/9

Ministry of Defence (MOD – UK), **23**, **99–101**, **121**, 1/71, 5/83

Ministry of Overseas Development (Dept. for International Development – UK), **25**, 5/6

Minority Rights Group (NGO), 1/64, 2/31

missile technology control regime, **40**

missionaries, **1–2**, 1/5

Moffatt & Nichol Engineers, 5/80

Mooring/anchorage fees, **31**, **86–7**, **94**, 2/55

Morocco, 4/35

Moulinié & Co., **21–2**

Mowlem & Co. PLC, **35**

Mozambique, **15**, 2/37

multilateral agreements, *see* conventions

Nairobi Convention, *see* conventions

Napoleonic Wars, **15**, 1/2

National Oceanic and Atmospheric Administration (NOAA – USA), 4/20, 5/53

National Security Agency (NSA – USA), **44**, 4/9

natural heritage, **51**, **68**
  see also conventions, world cultural and natural heritage

Natural Resources Management Plan Diego Garcia (NRMP), **12**, **40**, **53**, **55–7**, 1/75, 3/60, 5/21, 5/26–7, 5/41, 5/45–7, 5/59–60, 5/64

New Zealand, **44**, 2/16, 4/7

non-governmental organizations (NGOs), **56**, **65**, 1/64, 1/69, 4/25, 4/33, 4/36, 6/28

non-self-governing territories, **2–3**, **65**, **136**, 1/14, 2/54, 6/29

North Atlantic Treaty Organization (NATO), **49**, 1/68, 1/74

Norway, 1/77

nuclear material/radiation/weapons, **5**, **37–41**, **58**, 1/57, 3/27–46, 3/50–3, 5/66

nuclear non-proliferation, *see* conventions

nuclear-powered ships, **37–8**, **40**, 1/57, 3/18, 5/82

nuclear test ban, *see* conventions

ocean acidification, 5/73

Oceaneering International, **52**, 5/16

Ocean Engineering and Energy Systems International (OCEES), 5/84

Ocean Thermal Energy Conversion Systems Ltd. (OTECS), 5/84

oil and mineral exploration, **19**, **22**, **87**, **95**

Oman, **39**, **41**

orders-in-council (BIOT), **4–5**, **27**, **30–2**, **122–4**, **130–2**, **135**, 1/22, 6/6

ordinances (BIOT), **4**, **17**, **23–4**, **30**, **124**, **126–7**, **139**, **141**, **143**, 1/60, 2/10, 5/3

Subject Index

ordnance, *see* ammunition
Organization for African Unity (OAU, African Union), **38**
Ottawa Convention, *see* conventions
ozone layer treaties, *see* conventions

Panama, **56**, 5/54
Papahanaumokuakea marine reserve, (USA), 6/49
Pelindaba Treaty, *see* conventions
Pentagon (USA), **2**, **5**, 16, **18**, 24, 37, 39–41, **59**, **124**, 1/35, 1/69, 2/26, 3/2, 3/11, 3/50, 5/38, 5/42, 5/56
Penta-Ocean Construction Co., **35**, 3/7
Permanent Court of Arbitration (PCA), **4**
permanent inhabitants, **16–17**, **19**, **30–3**, 2/21, 2/27
Peros Banhos Islands, **20–1**, **122**, **124**, **126**, 1/4, 1/13, 2/12, 2/24, 2/55
Pew Oceans Commission, 6/55
Philippines, **37**, **56**, 3/13, 3/16, 5/54
Pitcairn Islands, **59**, 1/31, 2/16, 2/62
POLARIS Sales Agreement (USA/UK), **5–7**, **25**, 1/34
'political question' doctrine, *see* justiciability
pollution, *see* fuel spills
Portugal, **1**, 1/1
Posford Haskoning, 2/60, 6/10
'prepositioning', **1**, **11**, 3/20
'prerogative rights' doctrine, **29**, **31**, **63**, **122**, **131–2**, **135–6**, 2/47, 2/62–3, 6/6–7
prison ships, **45**, **48**, **133**, 5/82
'prizes', **15**, **79**
public trust concept, **67–8**, 6/52–5
Pueblo incident, **44**, 4/13
Puerto Rico, **56**, 5/54

Qatar, **41**

radiation, *see* nuclear material
Rambler Bay, **11–12**
Ramsar Convention, *see* conventions

Raymond International, Brown & Root (RBRM), **35**, 3/8, 3/10
Redress (NGO), 4/36
reef-blasting, **52**, **54**, **108**, 5/16, 5/35, 5/44
rendition flights, **47–8**, **133**, 4/22, 4/28, 4/33–6, 5/61, 6/1
Reprieve (NGO), 4/25, 4/27, 4/36
repugnancy to English law, **132**, **134**
resettlement, **3**, **18–23**, **25–7**, **30–2**, **59**, **64**, **81**, **126**, **128–30**, **139**, **142**, 2/26, 2/33, 2/60, 2/63, 6/11
Réunion, 5/74, 6/39
Rockall Island, 2/58
rocks in the sea, **17**, **32**, 2/57–8
Rome Statute, *see* International Criminal Court
Rose atoll, 6/29
Royal Naval Birdwatching Society, 5/6
Royal Ordnance PLC, 5/33
Royal Society of London, **2**, 1/9, 5/2, 5/73
Russia, **1**, **37**, **44**, **125**, 3/24, 3/48–9, 4/14

'sacred trust' concept, **20**, **65**, **140**, 6/26, 6/52
Salomon Islands, **20–1**, **122**, **124**, **126**, 1/13, 2/12, 2/24, 2/55, 4/50
satellites, *see* space tracking
Scaevola, 5/29
'Seabees' (naval construction battalions), **23**, **35**, **52**, 3/5, 5/16, 5/44
sea-level rise, **59**, **128**, 5/74–6
seawater air-conditioning (SWAC), 5/84
secret side-notes on Diego Garcia (UK/USA), **5–8**, **45**, **81–3**, 1/34, 1/37
seismographic monitoring, **45–6**, 4/18, 4/21, 4/47–8, 4/54
self-determination/self-government, **2–3**, **16**, **20**, **65**, 1/14, 1/30, 2/54, 6/29, 6/31
Seychelles, **3–5**, **9**, **16–17**, **19–22**, **24–5**, **71**, **83**, **88**, **96**, **122–3**, 1/9, 1/27, 1/30, 1/33, 1/51–2, 2/24, 6/39

signals intelligence (SIGINT), **43–4**, 4/6
Singapore, **36**, **53**, 3/3
SIOFA, *see* conventions
slavery, **15**, **66**, 2/3–4
Smithsonian Institution, **2**, 4/19
Société Huilière de Diego et Peros, 1/4
Somalia, **59**
sonar (sound navigation and ranging), **49**, **58**, 4/47–9, 4/51, 4/53
sovereign immunity, **10–11**, **29**, **87**, **95**, 1/66, 2/41, 4/13
sovereignty issues, **8–9**, **38–9**, **66–7**, **70**, **87**, **95**, 1/33, 1/38, 1/47–9, 1/52, 6/26, 6/34, 6/53–4
Soviet naval power, *see* Russia
space tracking, **4**, **44–5**, **81**, **83**, **118–21**, 1/27, 4/16
Spain, **1**, 1/1, 1/47
Sri Lanka, **15**, **37–8**, 1/48, 2/4, 5/76
St. Helena Island, 5/69
Stockholm International Peace Research Institute (SIPRI), 3/23, 3/46
Strategic Arms Reduction Treaty (START-1, Russia/USA), **40**, 3/48
'strategic islands' concept, **1–2**, **61**, 1/11
submarines, **1**, **5**, **37**, **39–40**, **44**, **49**, **59**, 3/18, 3/41, 3/44, 3/47, 3/49, 3/52, 4/7, 4/42, 4/48, 5/82
Supreme Court (USA), **29**

Taiwan, **35**, **52**, 5/44
tangan-tangan, *see Leucaena leucocephala*
'Tarzans', **17**
taxation (BIOT), **71**, **73–5**, **83**, 3/14
territorial waters, **9–11**, **47–8**, **51–2**, **61**, **87**, **95**, 1/47–9, 1/53, 1/57, 1/72
terrorist suspects, **31**, **46–8**, **63**, **140**, 4/34, 4/36, 6/1
tortoises, **2**, 1/9
torture, **10**, **46**, **64**, **133**, 4/22, 4/35–6, 4/38
treaties, *see* conventions
Tristan da Cunha Island, 5/69

Tromelin Island, 1/38, 1/48
tsunami, **45–6**, 4/19–21

'UKUSA Agreement', **44**, 4/7
underwater noise pollution/signalling, **44**, **49**, 4/17, 4/47–9, 4/51
United Nations (UN) Ad Hoc Committee on the Indian Ocean, **38**
United Nations Charter, **2**, **16**, **20**, **24**, **65**, **125**, **140**, **190**, 1/30, 2/54
United Nations Convention on the Law of the Sea (UNCLOS), *see* conventions
United Nations Educational, Scientific and Cultural Organization (UNESCO), 5/4–5
United Nations Environment Programme (UNEP), 1/43, 4/7, 4/43, 5/78
United Nations General Assembly (UNGA), **2**, **4**, **38**, 1/14–15, 1/26, 1/39–40, 3/30–1, 3/34, 6/23, 6/29, 6/36
United Nations Human Rights Committee (UNHRC), 1/64, 4/34
uti possidetis doctrine, **4**, 1/25

warships, **1**, **10–11**, **37–40**, **45**, **48**, **85–6**, **92–4**, 1/57, 1/59, 3/18, 3/44, 4/13, 5/82
Washington Post, **25**, 2/28
water pollution/quality, **56**, **58**, **107–8**, 3/53, 5/50, 5/55, 5/65
Westfall Act (USA), 2/40
wetlands, *see* conventions
whales, **49**, 4/48–50, 4/52–3
World Conservation Union (IUCN), 4/43, 4/51, 5/46, 5/74
world heritage, *see* conventions

'yachties', **31**, 2/55

'zone of peace', **38**, 3/30–1